Fungal Infections in Intensive Care Medicine

Fungal Infections in Intensive Care Medicine

Guest Editor

Ayman O. Soubani

MDPI

Basel • Beijing • Wuhan • Barcelona • Belgrade • Novi Sad • Cluj • Manchester

Guest Editor
Ayman O. Soubani
Division of Pulmonary,
Critical Care and Sleep
Medicine
Wayne State University School
of Medicine
Detroit
USA

Editorial Office
MDPI AG
Grosspeteranlage 5
4052 Basel, Switzerland

This is a reprint of the Special Issue, published open access by the journal *Journal of Fungi* (ISSN 2309-608X), freely accessible at: https://www.mdpi.com/journal/jof/special_issues/HH866PT4JT.

For citation purposes, cite each article independently as indicated on the article page online and as indicated below:

Lastname, A.A.; Lastname, B.B. Article Title. *Journal Name* **Year**, *Volume Number*, Page Range.

ISBN 978-3-7258-7348-7 (Hbk)
ISBN 978-3-7258-7349-4 (PDF)
https://doi.org/10.3390/books978-3-7258-7349-4

Contents

About the Editor

Ayman O. Soubani

Ayman O. Soubani, MD, is Professor of Medicine and Chief of the Division of Pulmonary, Critical Care, and Sleep Medicine at Wayne State University School of Medicine in Detroit, Michigan, USA. He also holds an appointment as an Honorary Professor at the University of Jordan in Amman, Jordan. His clinical and scholarly work focuses on pulmonary aspergillosis and the pulmonary and critical care management of patients with cancer. Dr. Soubani has authored numerous peer-reviewed publications and edited books on these topics, and he serves on the Editorial Boards of several scientific journals.

Journal of *Fungi*

MDPI

Editorial

An Evolving Challenge: Fungal Infections in the Contemporary ICU

Ayman O. Soubani

Division of Pulmonary, Critical Care and Sleep Medicine, Wayne State University School of Medicine, Detroit, MI 48201, USA; asoubani@med.wayne.edu

Fungal infections have become an integral and increasingly complex component of modern intensive care medicine. Improvements in critical care survival have expanded the population of patients exposed to prolonged mechanical ventilation, broad-spectrum antimicrobial therapy, central venous access, extracorporeal organ support, and immunomodulatory agents, all of which contribute to profound alterations in host immunity. Fungal infections in critically ill patients are common and are no longer limited to those who are immunocompromised. They are also associated with excess mortality, prolonged length of stay, and substantial healthcare utilization, particularly when diagnosis and initiation of appropriate antifungal therapy are delayed.

Invasive candidiasis remains the most frequently encountered fungal infection in critically ill patients and constitutes the majority of fungal bloodstream infections in the ICU. While *Candida albicans* historically predominated, there has been a sustained global shift toward non-albicans species, including *Candida glabrata*, *Candida parapsilosis*, *Candida tropicalis*, and, more recently, the multidrug-resistant *Candida auris*. This epidemiologic evolution has significant therapeutic and stewardship implications, given species-specific differences in virulence, antifungal susceptibility, and propensity for nosocomial transmission. It is currently recommended to use an echinocandin as first-line empiric therapy for most ICU patients with suspected invasive candidiasis, while highlighting the importance of early de-escalation based on species identification and susceptibility testing [1].

In parallel, invasive aspergillosis has emerged as a major cause of morbidity and mortality in the ICU beyond its traditional association with neutropenia and hematologic malignancy. Critically ill patients with acute respiratory distress syndrome (ARDS), chronic obstructive pulmonary disease, liver failure, or prolonged corticosteroid exposure are now recognized as being at substantial risk [2]. Influenza-associated pulmonary aspergillosis (IAPA) and COVID-19-associated pulmonary aspergillosis (CAPA) have been highlighted by recent respiratory viral pandemics and reflect the role of severe respiratory viral infections and critical illness in predisposing to invasive aspergillosis [3]. Mortality in these populations remains high and is compounded by diagnostic uncertainty and delayed initiation of mold-active antifungal therapy.

Other rare fungal infections that may be associated with critical illness and require ICU care include mucomycosis in severely immunocompromised patients and those with diabetic ketoacidosis. Pneumocystis jirovecii in patients with HIV and other immunosuppressive conditions is another fungal infection that may be associated with acute respiratory failure and the need for invasive mechanical ventilation. Endemic fungi such as histoplasma and blasomyces are rare but may also be encountered.

Reported incidence rates of invasive fungal infections in the ICU vary widely according to case mix, surveillance intensity, and geographic region, but contemporary studies estimate that approximately 3–10% of critically ill patients develop these types of serious infections, with higher rates in selected populations, such as those with severe viral pneumonia or complicated intra-abdominal infections [4]. Across etiologies, mortality remains unacceptably high and is consistently linked to delays in appropriate antifungal treatment, reinforcing guideline recommendations for early therapy in high-risk patients while avoiding indiscriminate antifungal use [1,5].

The timely diagnosis of fungal infections in critically ill patients remains a central challenge. Clinical features are nonspecific and frequently indistinguishable from bacterial sepsis or sterile inflammatory syndromes. Conventional culture-based diagnostics lack sensitivity and are often delayed, while differentiation between colonization and invasive disease—particularly for *Candida* in respiratory or urinary specimens and *Aspergillus* in airway samples—remains problematic. Multiple diagnostic criteria have been suggested for the diagnosis of IFI in critically ill patients that endorse the use of non-culture diagnostics, including β-D-glucan, galactomannan, and molecular assays, as adjunctive tools, but acknowledge important limitations in ICU populations, including false positivity related to renal replacement therapy, antibiotic exposure, mucosal injury, and critical illness itself [6].

Therapeutic management is further complicated by altered pharmacokinetics in critical illness, organ dysfunction, extracorporeal support modalities, and clinically significant drug–drug interactions. Triazole antifungals require careful attention to absorption, metabolism, and therapeutic drug monitoring, while emerging resistance among *Candida* spp. and azole-resistant *Aspergillus* increasingly constrain treatment options. These challenges underscore the importance of ICU-specific antifungal stewardship strategies that integrate local epidemiology, rapid diagnostics, and individualized pharmacologic optimization.

Against this backdrop, future advances in the management of ICU-associated fungal infections will depend on improved early diagnostic strategies, refined risk stratification, and the development of novel antifungal agents with activity against resistant pathogens. The integration of predictive models, host immune biomarkers, and real-time surveillance data may allow for a transition from empiric to precision-guided antifungal therapy.

This Special Issue of the *Journal of Fungi* presents a synopsis of fungal infections in intensive care medicine, with original research and review articles that primarily focus on the two main severe fungal infections: candida and Aspergillus. Research papers on candida discuss resistance patterns, infection in cardiothoracic ICU, and the association with severe COVID infection. Another paper provides insight into the epidemiology and outcomes of invasive aspergillosis in patients with liver failure. The review papers discuss the state of the art in terms of the risk factors, diagnosis and management of these serious infections in critically ill patients.

Conflicts of Interest: The authors declare no conflicts of interest.

References

1. Pappas, P.G.; Kauffman, C.A.; Andes, D.R.; Clancy, C.J.; Marr, K.A.; Ostrosky-Zeichner, L.; Reboli, A.C.; Schuster, M.G.; Vazquez, J.A.; Walsh, T.J.; et al. Clinical Practice Guideline for the Management of Candidiasis: 2016 Update by the Infectious Diseases Society of America. *Clin. Infect. Dis.* **2016**, *62*, e1–e50. [CrossRef] [PubMed] [PubMed Central]
2. Elkhapery, A.; Fatima, M.; Soubani, A.O. Emerging Risk Factors for Invasive Pulmonary Aspergillosis: A Narrative Review. *J. Fungi* **2025**, *11*, 555. [CrossRef] [PubMed] [PubMed Central]
3. Pervaiz, A.; Soubani, A.O. Virus-associated pulmonary aspergillosis: A rising challenge in respiratory infections. *Am. J. Med. Sci.* 2025, *epub ahead of print*. [CrossRef] [PubMed]
4. Trelles, M.; Murillo, J.; Fuenmayor-González, L.; Yu-Liu, Y.; Alexander-León, H.; Acebo, J.; Cuicapuza, D.; Morales, M.; Peña, C.; García-Aguilera, M.F. Prevalence of invasive fungal infection in critically Ill patients: A systematic review and meta-analysis. *BMC Infect. Dis.* **2025**, *25*, 896. [CrossRef] [PubMed] [PubMed Central]

https://doi.org/10.3390/jof12030197

5. Patterson, T.F.; Thompson, G.R., 3rd; Denning, D.W.; Fishman, J.A.; Hadley, S.; Herbrecht, R.; Kontoyiannis, D.P.; Marr, K.A.; Morrison, V.A.; Nguyen, M.H.; et al. Practice Guidelines for the Diagnosis and Management of Aspergillosis: 2016 Update by the Infectious Diseases Society of America. *Clin. Infect. Dis.* **2016**, *63*, e1–e60. [CrossRef] [PubMed] [PubMed Central]
6. Bassetti, M.; Giacobbe, D.R.; Agvald-Ohman, C.; Akova, M.; Alastruey-Izquierdo, A.; Arikan-Akdagli, S.; Azoulay, E.; Blot, S.; Cornely, O.A.; Cuenca-Estrella, M.; et al. Invasive Fungal Diseases in Adult Patients in Intensive Care Unit (FUNDICU): 2024 consensus definitions from ESGCIP, EFISG, ESICM, ECMM, MSGERC, ISAC, and ISHAM. *Intensive Care Med.* **2024**, *50*, 502–515. [CrossRef] [PubMed] [PubMed Central]

 https://doi.org/10.3390/jof12030197

Journal of *Fungi*

MDPI

Article

Candida albicans Horizontal Transmission in COVID-19 Patients Hospitalized in Intensive Care Unit

Magdalena Skóra [1,*], Katharina Rosam [2], Magdalena Namysł [3,4], Anna Sepioło [3], Mateusz Gajda [1], Justyna Jędras [5], Paweł Krzyściak [1], Joanna Zorska [6,7], Jerzy Wordliczek [8], Piotr B. Heczko [1], Reinhard Würzner [2,*], Michaela Lackner [2] and Jadwiga Wójkowska-Mach [1]

1 Chair of Microbiology, Faculty of Medicine, Jagiellonian University Medical College, Czysta 18 Street, 31-121 Krakow, Poland; mateusz14.gajda@uj.edu.pl (M.G.); pawel.krzysciak@uj.edu.pl (P.K.); piotr.heczko@uj.edu.pl (P.B.H.); jadwiga.wojkowska-mach@uj.edu.pl (J.W.-M.)
2 Institute for Hygiene and Medical Microbiology, Medical University of Innsbruck, Schöpfstraße 41, 6020 Innsbruck, Austria; katharina.rosam@i-med.ac.at (K.R.); michaela.lackner@i-med.ac.at (M.L.)
3 Department of Microbiology, University Hospital in Krakow, Jakubowskiego Street 2, 30-688 Krakow, Poland; mnamysl@su.krakow.pl (M.N.); asepiolo@su.krakow.pl (A.S.)
4 Department of Microbiology, Faculty of Pharmacy, Jagiellonian University Medical College, Medyczna Street 9, 30-688 Krakow, Poland
5 Hematology Clinic, University Hospital in Krakow, Jakubowskiego Street 2, 30-688 Krakow, Poland; jbjedras@gmail.com
6 Center for Innovative Medical Education, Jagiellonian University Medical College, Medyczna 7 Street, 30-688 Krakow, Poland; joanna.zorska@uj.edu.pl
7 Intensive Care Unit, University Hospital in Krakow, Macieja Jakubowskiego 2 Street, 30-688 Krakow, Poland
8 Interdisciplinary Intensive Care Clinic, Jagiellonian University Medical College, Macieja Jakubowskiego 2 Street, 30-688 Krakow, Poland; j.wordliczek@uj.edu.pl
* Correspondence: magdalena.skora@uj.edu.pl (M.S.); reinhard.wuerzner@i-med.ac.at (R.W.)

Abstract: Background: Invasive candidiasis is a predominant mycosis in hospitalized patients, and *Candida albicans* is the species most often responsible for this infection. Most candidiasis cases originate from endogenous mycobiota; therefore, *Candida* strains can easily be transferred among hospital patients and personnel. The aim of this study was to assess the possible horizontal transmission of *C. albicans* in patients with severe COVID-19 infection requiring hospitalization in the intensive care unit. Methods: In total, 59 *C. albicans* strains from 36 patients were collected from blood and lower-respiratory samples. The strains were genotyped using the RAPD method with the OPA-18 primer (5′-AGCTGACCGT-3′). Antifungal susceptibility testing was performed for amphotericin B (AMB), fluconazole (FCZ), voriconazole (VCZ), and anidulafungin (ANF) using the EUCAST method. Results: *C. albicans* strains were separated into 13 different groups according to their RAPD pattern. Two predominant clonal clusters of 17 strains isolated from 12 patients and 12 strains from 7 patients were identified, followed by clusters with 6, 4, and 8 strains isolated from 5, 4, and 3 patients, respectively. *C. albicans* strains were sensitive to AMB, FCZ, VCZ, and ANF, and antifungal susceptibility profiles were similar in all genetic clusters. Conclusions: Our study indicates that *C. albicans* strains can spread horizontally. The routes of transmission for strains in the ward have not been explained due to there being insufficient data. The transmission could have been caused by the unintentional spread of pathogens by medical personnel.

Keywords: *Candida albicans*; horizontal transmission; outbreak; intensive care unit; SARS-CoV-2; molecular epidemiology; COVID-19 pandemic; genotyping

1. Introduction

Candida is a predominant human fungal pathogen that causes a wide spectrum of infections, ranging from mucocutaneous to deep-seated. The species *Candida albicans* is still the most frequently isolated from clinical samples and a leading cause of candidiasis;

however, an increase in the clinical importance of other species, including those intrinsically less susceptible or resistant to antifungal drugs, has been observed over many years [1,2]. It is estimated that non-*albicans* species may constitute more than 50% of the etiological factors of candidemia [3,4]. In the last two decades, *Candida auris* in particular has attracted particular attention and concern. It is known for its multidrug resistance, resistance to disinfectants, and ability to cause healthcare-associated outbreaks due to its long survival period in the environment and the possibility of horizontal transmission by, e.g., healthcare workers [5].

The source of infection, regardless of the form of candidiasis and its severity, is usually the patient's natural mycobiota colonizing the skin and mucous membranes. Disturbances to immune and hormonal homeostasis and other components of the microbiota, as well as an impaired skin barrier and gastrointestinal mucous membrane breakage, may lead to the intensive colonization of endogenous *Candida* spp. and their translocation to other tissues, including physiologically sterile sites [2]. Life-threatening invasive infections mainly affect patients undergoing immunosuppressive therapy, those with long-term use of corticosteroids or antibiotics, and those with long-term vascular catheters. Critically ill patients in intensive care units (ICUs) are at high risk for developing invasive candidiasis due to the accumulation of multiple predisposing factors [6–9]. Thus, the diagnosis of infection in this group of patients is difficult and complex. These patients often do not meet the criteria for an invasive fungal disease according to guidelines developed by the European Organization for Research and Treatment of Cancer (EORTC) and the Mycoses Study Group Education and Research Consortium (MSGERC) [10]. Therefore, they require separate, thorough diagnostics and the individual interpretation of test results [11–13].

During the global COVID-19 pandemic, it was observed that infection with, and a severe course of, the SARS-CoV-2 virus, leading to the subsequent need for intensive hospital care, also represent a factor favoring fungal infections. In COVID-19 patients, invasive candidiasis was among the most commonly reported mycoses, along with pulmonary aspergillosis and mucormycosis [14,15]. The increased susceptibility of COVID-19 patients to fungal infections and the higher incidence of candidemia in critically ill COVID-19 patients compared to non-COVID-19 patients are due to the effects of the virus dysregulating the host's immune system and the medical procedures for treating coronavirus disease. These factors, among others, could lead to a reduced ability to defend against fungal pathogens [14,16]. In Poland, the COVID-19 pandemic has had a significant impact on the health service and medical care. Temporary units and wards were created to treat patients requiring hospitalization due to coronavirus disease. In particular, in the first period of the pandemic, the organization of medical care was quite chaotic, and access to various procedures and personal protective equipment (PPE) for healthcare workers was difficult. These factors influenced the quality of medical care and the comfort of work for medical staff [17]. Newly introduced procedures, limitations in resources, and the discomfort associated with the long-term use of PEE could have intensified the horizontal transmission of microorganisms in the hospital environment and increased the risk of infections with hospital strains. Horizontal transfer between patients admitted in COVID-19 units was described for bacterial pathogens [18–20]. *C. albicans* and other *Candida* species have been reported to cause nosocomial outbreaks in the past, although before the era of *C. auris*, they were considered rather rare [21–23]. Therefore, the question arose as to whether, during the COVID-19 pandemic, there was horizontal transmission of the most common *Candida* species—*C. albicans*—in wards where patients with severe coronavirus disease were hospitalized.

Objective

The current publication aimed to assess the possible horizontal transmission of *C. albicans* in COVID-19-positive ICU patients during the pandemic period. Due to the retrospective nature of this analysis and insufficient clinical data, this study did not examine candidiasis and the distinction between *C. albicans* infection and colonization, but only strain transmission between patients staying on the same ward. The study included only strains deposited during routine microbiological diagnostics performed during the patients' hospitalization. Due to the sanitary rules introduced in the COVID-19 pandemic, it was not possible to carry out additional tests, and no environmental or medical personnel studies were performed. The current analysis was limited to the species *C. albicans*, which turned out to be the predominant *Candida* species from clinical materials. Therefore, *C. albicans* isolates were genotyped, and their antifungal susceptibility profiles were compared.

2. Materials and Methods

2.1. Study Setting and Patient Population

The study covers the period between 1 May 2021 and 31 January 2022, during which 1614 patients (18,810 patient days, pds) with COVID-19, confirmed by SARS-CoV-2 PCR tests (COBAS 6800, Roche, Basel, Switzerland or in the CITO mode GeneXpert System, Cepheid, Sunnyvale, CA, USA), were hospitalized in the COVID-19-dedicated ICU of University Hospital in Krakow, Poland. Laboratory tests were performed depending on the patient's clinical condition.

2.2. Candida Detection and Isolation

Classic culture methods and serological tests were used to detect *Candida* spp. in clinical samples. In the analyzed period, mannan and anti-mannan antibody-detecting assays (Platelia Candida Ag Plus and Platelia *Candida* Ab Plus, Bio-Rad, Marnes-la-Coquette, France) were performed in 22 and 15 patients, with 28 and 17 tests, respectively. Beta-D-glucan (Fungitell, Associates of Cape Cod, Inc., East Falmouth, MA, USA) was tested in 37 patients with 51 tests. All serological tests were carried out in accordance with the manufacturers' instructions. Altogether, 238 patients had multiple blood cultures (BacT/ALERT 3D 480, bioMerieux, Marcy-l'Etoile, France; BacT/ALERT VIRTUO CLINIC, bioMerieux, Marcy-l'Etoile, France) (1047 cultures in total), and 214 underwent fungal cultures of materials collected from the lower respiratory tract (LRT) (405 tests in total) including bronchoalveolar lavage (BAL) and other materials (bronchial lavage, tracheal aspirate, and secretion from the bronchial tree), called, for the purposes of this publication, non-bronchoscopic lavage (NBL). *Candida* strains were isolated on Sabouraud glucose agar with gentamicin and chloramphenicol (OXOID) at 25 °C and 35 °C. Species identification was performed with matrix-associated laser desorption ionization–time-of-flight mass spectrometry (MALDI-TOF MS) VITEK®MS (bioMerieux, Marcy-l'Etoile, France) (ver. V3.2). The strains were stored at minus 70 °C for further research. Before genotyping the species, identification was confirmed via Sanger sequencing, using ITS primers.

2.3. Candida Albicans Genotyping

Genomic DNA was extracted from freshly grown colonies on Sabouraud glucose agar with cetyltrimethylammonium bromide (CTAB) according to the method described by Lackner et al. [24]. The strains were genotyped using the random amplified polymorphic DNA (RAPD) method with the OPA-18 primer (5′-AGCTGACCGT-3′). The PCR mixture contained 12.5 µL mastermix (Kappa2G Fast Hotstart, Merck KG, Darmstadt, Germany), 0.5 µM primer, and 30 ng of DNA template in a total reaction volume of 25 µL. The procedure included 38 cycles of 60 s at 94 °C and 60 s at 36 °C and two minutes at 72 °C according to Bautista-Munoz et al. 2003 [25]. The samples were separated on a 1.2% agarose gel (Biozym Scientific GmbH, Hessisch Oldendorf, Germany) at 170 V for 90 min. The band patterns were compared visually in sets; a difference of ≥3 was considered as indicating different genotypes.

2.4. Candida Albicans Antifungal Susceptibility Testing

The antifungal susceptibility testing (AFST) of *C. albicans* isolates was performed according to the European Committee on Antimicrobial Susceptibility Testing (EUCAST) method for yeasts [26]. Minimal inhibitory concentration (MIC) values were obtained for the amphotericin B (AMB) (Pol-Aura, Morąg, Poland), fluconazole (FCZ) (Pol-Aura, Poland), voriconazole (VCZ) (Pol-Aura, Poland), and anidulafungin (ANF) (Merck, Darmstadt, Germany). RPMI 1640 medium with L-glutamine and without sodium bicarbonate (Merck, Germany) supplemented with 2% glucose (Chempur, Piekary Śląskie, Poland) and buffered to pH 7 with 4-morpholinepropanesulfonic acid (MOPS; 0.165 mol/L, Fluorochem, Pune, India) was used as a culture medium. Dimethyl sulfoxide (DMSO) (Chempur, Poland) served as a solvent and diluent for drugs. The studies were performed using flat-bottom polystyrene 96-well microdilution plates (Nest, Wuxi, China). The results were read with a microdilution plate reader (Tecan Sunrise, Männedorf, Switzerland), using a wavelength of 530 nm, and interpreted according to the breakpoints determined by the EUCAST [27].

3. Results

During the analyzed period, 104 *Candida* strains from 53 COVID-19 patients hospitalized in the temporary COVID-19-dedicated ICU of The University Hospital in Krakow, Poland, were isolated from blood samples (11 strains) and LRT samples (93 strains). The isolated species included *C. albicans* (56.73%), *C. glabrata* (15.38%), *C. tropicalis* (12.5%), *C. dubliniensis* (9.62%), *C. inconspicua* (2.88%), *C. parapsilosis* (1.92%), and *C. lusitaniae* (0.96%). *C. auris* was not detected.

In total, 59 *C. albicans* strains (4 blood isolates and 55 LRT isolates) from 36 patients were genotyped, and antifungal drug susceptibility profiles were compared (Tables 1–3). One strain did not survive the freezing procedure (strain no 73, Table 1) and was not included in the study. Patient characteristics and data on co-infection with other *Candida* species are included in Table 1. The OPA-18 primer gave RAPD patterns of 3->10 bands (Supplementary Information Figure S1). *C. albicans* strains were separated into 13 different groups according to their RAPD pattern (Table 2, Figure 1). The biggest group forming one clonal cluster (cluster no. 2) was found to contain 17 *C. albicans* strains and was linked to 12 patients. Another big *C. albicans* cluster (cluster no. 1) was found to contain 12 *C. albicans* strains from seven patients followed by clusters no. 5, no. 4, and no. 6, with 6, 4, and 8 strains isolated from five, four, and three patients, respectively. All other clusters were minor, with 1–4 isolates per RAPD pattern (Table 2).

Table 1. Characteristics of COVID-19 patients hospitalized in the intensive care unit with the isolation of *Candida albicans* from blood and lower respiratory tract samples.

Patient Characteristics							*C. albicans* Cultures						
Patient No	Sex	Age [Year]	LOS [Day]	LOS/ICU [Day]	CVC	MV	Strain No	Type of Sample	Sample Collection Date	RAPD Genetic Cluster	Other *Candida* Species Cultures (Type of Sample, Date of Collection)	Antifungal Treatment	Death
1	M	62	76	76	Y	Y	122/A	NBL	15 October 2021	5	*C. tropicalis* (NBL, 15 October and 19 November 2021) *	N	Y
2	M	35	9	9	Y	N	96	NBL	1 September 2021	2	N	N	N
3	M	71	32	32	Y	Y	6	NBL	12 May 2021	2	*C. glabrata*, *C. tropicalis* (NBL, 18 April 2021)	N	Y
							18	BAL	17 May 2021	2			
4	F	73	14	14	Y	Y	163	NBL	9 September 2021	2	N	Y (1-day FCZ)	Y
5	F	75	52	51	Y	Y	32	Blood	23 May 2021	2	*C. glabrata* (NBL, 12 April and 5 May 2021) *	Y (1-day VCZ)	Y
							33	Blood	21 May 2021	2			
6	M	60	13	13	Y	Y	28A	NBL	22 May 2021	2	*C. glabrata* (NBL, 22 May 2021) (urine, 22 May 2021) *	N	Y
7	M	81	2	2	Y	Y	138	NBL	2 November 2021	2	N	N	Y
8	M	71	27	27	Y	Y	23	NBL	19 May 2021	13	N	N	Y
9	F	82	18	12	Y	Y	147	BAL	4 November 2021	3	N	N	Y
10	M	69	7	2	Y	Y	67	NBL	19 June 2021	5	N	N	Y
11	F	63	23	23	Y	Y	117	BAL	8 October 2021	2	*C. lusitaniae* (NBL, 21 September and 24 September 2021) *	N	Y
							109A	BAL	21 September 2021	2			
							111B	BAL	24 September 2021	9			

Table 1. *Cont.*

Patient Characteristics							*C. albicans* Cultures						
Patient No	Sex	Age [Year]	LOS [Day]	LOS/ICU [Day]	CVC	MV	Strain No	Type of Sample	Sample Collection Date	RAPD Genetic Cluster	Other *Candida* Species Cultures (Type of Sample, Date of Collection)	Antifungal Treatment	Death
12	F	73	78	77	Y	Y	77	NBL	6 August 2021	6	N	Y (FCZ)	N
							85	NBL	24 August 2021	2			
							79A	BAL	12 August 2021	2			
							91A	NBL	30 August 2021	2			
13	F	51	42	22	Y	Y	45	NBL	29 May 2021	8	*C. glabrata* (urine, 13 May and 22 May and 29 May 2021)	Y (3-day CAS)	Y
14	F	72	6	6	Y	Y	161	NBL	9 November 2021	5	N	N	Y
15	F	80	21	21	Y	Y	123	NBL	21 October 2021	1	N	N	Y
							145	NBL	3 November 2021	1			
16	F	77	46	46	Y	Y	56	BAL	11 June 2021	8	N	N	N
17	F	75	15	15	Y	Y	116	NBL	7 October 2021	4	N	N	Y
18	M	32	34	x	N	N	10B	BAL	13 May 2021	11	N	Y (FCZ)	N
19	F	66	61	40	Y	Y	73	NBL	4 July 2021	Not tested	N	N	N
							74	NBL	8 July2021	2			
20	M	60	27	9	Y	Y	108	NBL	19 September 2021	5	N	Y (VCZ)	Y
							104B	NBL	15 September 2021	5			
21	M	31	10	9	Y	Y	136	NBL	1 November 2021	1	N	N	N
							146	NBL	3 November 2021	1			
							151	BAL	4 November 2021	1			
22	M	56	27	16	Y	Y	8B	NBL	12 May 2021	4	N	Y (FCZ)	Y
23	M	65	32	22	Y	Y	102	NBL	15 September 2021	8	N	N	Y
24	M	68	18	18	Y	Y	132	BAL	30 October 2021	4	N	N	N
25	M	68	24	22	Y	Y	89	NBL	26 August 2021	7	N	Y (FCZ)	Y
							93	NBL	31 August 2021	1			
							99	NBL	7 September 2021	2			
26	F	66	23	5	Y	Y	159/A	NBL	9 November 2021	4	*C.glabrata* (NBL, urine 9 November and 13 November 2021), *C. krusei* (NBL, urine 9 Nomvember 2021) (urine, 13 November 2024), *C.dubliniensis* (blood, NBL, 13 November 2024)	N	Y
27	F	29	24	12	Y	Y	153	BAL	4 November 2021	1	N	N	N
28	M	76	54	53	Y	Y	80	NBL	13 August 2021	6	N	Y (17-day CAS) (5-day FCZ)	Y
							81	Blood	11 August 2021	6			
							86	NBL	24 August 2021	6			
							95	NBL	31 August 2021	6			
							98	NBL	4 September 2021	6			
							101	NBL	8 September 2021	6			
29	M	66	99	25	Y	Y	97	NBL	2 September 2021	6	N	N	Y
30	M	58	11	4	Y	Y	134	NBL	31 October 2021	1	N	N	Y
							150	NBL	4 November 2021	2			
31	F	71	13	11	Y	Y	14	NBL	14 May 2021	7	N	N	Y
32	M	73	38	30	Y	Y	119	BAL	10 October 2021	1	N	N	Y
							124	NBL	22 October 2021	1			
33	F	33	29	17	Y	Y	43	NBL	28 May 2021	1	N	N	Y
							57A	BAL	9 June 2021	1			
34	M	82	95	63	Y	Y	112	NBL	26 September 2021	10	N	N	N
35	M	71	144	142	Y	Y	4	NBL	10 May 2021	12	N	Y (3-day CAS), (16-day FCZ)	N
							40	Blood	24 May 2021	5			
36	M	58	13	1	Y	Y	140	NBL	2 November 2021	2	N	N	Y

CVC—central venous catheter, LOS—length of hospital stay, LOS/ICU—length of stay in the intensive care unit, MV—mechanical ventilation, NBL—non-bronchoscopic lavage (lower respiratory tract samples other than bronchoalveolar lavage), FCZ—fluconazole, VCZ—voriconazole, CAS—caspofungin, N—no, Y—yes, * *C. albicans* was also isolated from the samples.

Table 2. *Candida albicans* genetic clusters in random amplified polymorphic DNA (RAPD) assay.

RAPD Genetic Cluster	*C. albicans* Strain No	Gel Picture No	Type of Clinical Sample	Sample Collection Date	Patient No	Hospitalization Date	Antifungal Treatment	Death
1	43	12	BAL	28 May 2021	33	20 May 2021	no	17 June 2021
	57A	15	BAL	9 June 2021				
	93	26	BAL	31 August 2021	25	16 August 2021	fluconazole (2 September 2021)	9 September 2021
	119	41	BAL	10 October 2021	32	27 September 2021	no	3 November 2021
	124	44	NBL	22 October 2021				
	123	43	BAL	21 October 2021	15	21 October 2021	no	11 November 2021
	145	50	NBL	3 November 2021				
	134	46	BAL	31 October 2021	30	25 October 2021	no	4 November 2021
	136	47	BAL	1 November 2021	21	30 October 2021	no	no
	146	51	BAL	3 November 2021				
	151	54	BAL	4 November 2021				
	153	55	BAL	4 November 2021	27	30 October 2021	no	no
2	33	10	Blood	21 May 2021	5	05.04.2021	voriconazole (26 May 2021)	26 May 2021
	32	9	Blood	23 May 2021				
	6	2	NBL	12 May 2021	3	17.04.2021	no	19 May 2021
	18	6	BAL	17 May 2021				
	28A	8	BAL	22 May 2021	6	13 May 2021	no	26 May 2021
	74	17	NBL	8 July 2021	19	25 June 2021	no	no
	79A	19	BAL	12 August 2021	12	5 August 2021	fluconazole	no
	85	22	BAL	24 August 2021				
	91A	25	BAL	30 August 2021				
	99	31	BAL	7 September 2021	25	16 August 2021	fluconazole	9 September 2021
	96	28	NBL	1 September 2021	2	1 September 2021	no	no
	109A	36	BAL	21 September 2021	11	20 September 2021	no	13 October 2021
	117	40	BAL	8 October 2021				
	140	49	BAL	2 November 2021	36	22 October 2021	no	3 November 2021
	150	53	BAL	4 November 2021	30	25 October 2021	no	4 November 2021
	163	58	BAL	9 November 2021	4	29 October 2021	fluconazole (11 November 2021)	12 November 2021
	138	48	BAL	2 November 2021	7	1 November 2021	no	3 November 2021
3	147	52	BAL	4 November 2021	9	23 April 2021	no	11 May 2021
4	8B	3	NBL	12 May 2021	22	22 April 2021	fluconazole (8 May 2021)	18 May 2021
	116	39	BAL	7 October 2021	17	26 September 2021	no	11 October 2021
	159/A	56	BAL	9 November 2021	26	23 October 2021	no	9 November 2021
	132	45	BAL	30 October 2021	24	30 October 2021	no	no
5	40	11	Blood	24 May 2021	35	4 April 2021	caspofungin (26 May–28 May 2021), fluconazole (28 May–12 June 2021)	no
	67	16	BAL	19 June 2021	10	13 June 2021	no	19 June 2021
	104B	34	BAL	15 September 2021	20	29 August 2021	voriconazole (22 September 2021)	24 September 2021
	108	35	BAL	19 September 2021				
	122/A	42	BAL	15 October 2021	1	3 October 2021	no	18 December 2021
	161	57	BAL	9 November 2021	14	4 November 2021	no	10 November 2021
6	97	29	BAL	2 September 2021	29	14 June 2021	no	20 September 2021
	77	18	BAL	6 August 2021	12	5 August 2021	fluconazole (14–18 August 2021)	no
	81	21	Blood	11 August 2021	28	5 August 2021	caspofungin (16 August–22 August 2021), fluconazole (29 August–2 September 2021)	28 September 2021
	80	20	BAL	13 August 2021				
	86	23	BAL	24 August 2021				
	95	27	BAL	31 August 2021				
	98	30	BAL	4 September 2021				
	101	32	NBL	8 September 2021				
7	14	5	BAL	14 May 2021	31	1 May 2021	no	14 May 2021
	89	24	BAL	26 August 2021	25	16 August 2021	fluconazole (2 September 2021)	9 September 2021
8	45	13	BAL	29 May 2021	13	24 April 2021	caspofungin (2 June–4 June 2021)	4 June 2021
	56	14	BAL	11 June 2021	16	5 June 2021	no	no
	102	33	BAL	15 September 2021	23	17 August 2021	no	18 September 2021
9	111B	37	BAL	24 September 2021	11	20 September 2021	no	13 October 2021
10	112	38	BAL	26 September 2021	34	12 September 2021	no	no

Table 2. *Cont.*

RAPD Genetic [illegible]	*C. albicans* Strain [illegible]	Gel [illegible]	Type of Clinical [illegible]	[illegible]	[illegible]	[illegible]	[illegible]	Death
11	10B	4	BAL	13 May 2021	18	30 April 2021	fluconazole	no
12	4	1	NBL	10 May 2021	35	4 April 2021	caspofungin (26–28 May 2021), fluconazole (28 May–12 June)	no
13	23	7	BAL	19 May 2021	8	1 May 2021	no	27 May 2021

BAL—bronchoalveolar lavage, NBL—non-bronchoscopic lavage (samples from lower respiratory tract other than bronchoalveolar lavage).

Table 3. Minimal inhibitory concentration (MIC) values [mg/L] obtained for the most numerous *Candida albicans* clusters.

C. albicans RAPD Cluster Number	AMB			FCZ			VCZ			ANF		
	MIC Range	MIC_{50}	MIC_{90}	MIC Range	MIC_{50}	MIC_{90}	MIC Range	MIC_{50}	MIC_{90}	MIC Range	MIC_{50}	MIC_{90}
1	0.012–0.25	0.125	0.125	≤0.125–0.25	0.25	0.25	≤0.008–0.016	≤0.008	0.016	≤0.008	≤0.008	≤0.008
2	0.06–0.25	0.125	0.125	≤0.125–0.5	0.25	0.5	≤0.008–0.031	0.016	0.03	≤0.008–0.031	≤0.008	0.016
4	0.06–0.25	0.125	0.25	0.125–0.25	0.25	0.25	≤0.008	≤0.008	≤0.008	≤0.008–0.016	≤0.008	0.016
5	0.06–0.125	0.125	0.125	≤0.125–0.5	0.25	0.5	≤0.008–0.031	≤0.008	0.03	≤0.008–0.016	≤0.008	0.016
6	0.06–0.125	0.125	0.125	≤0.125–0.25	≤0.125	0.25	≤0.008–0.016	≤0.008	0.016	≤0.008–0.016	≤0.008	0.016
7	0.125	0.125	0.125	≤0.125–0.25	≤0.125	0.25	≤0.008–0.031	≤0.008	0.03	≤0.008	≤0.008	≤0.008
8	0.125	0.125	0.125	0.25	0.25	0.25	0.016–0.031	0.016	0.03	≤0.008	≤0.008	≤0.008

AMB—amphotericin B, FCZ—fluconazole, VCZ—voriconazole, ANF—anidulafungin, MIC—minimal inhibitory concentration, MIC_{50}—concentration of antifungal drug able to inhibit 50% of strains tested, MIC_{90}—concentration of antifungal drug able to inhibit 90% of strains tested.

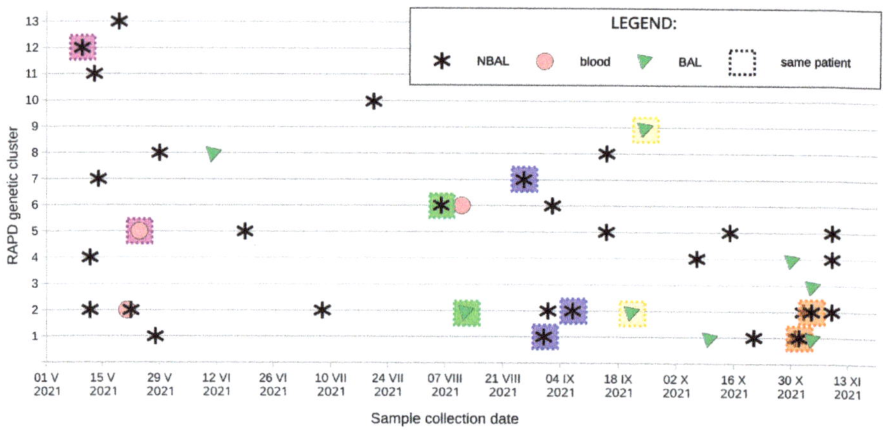

Figure 1. Isolation of *Candida albicans* strains belonging to different genotypes during the analyzed period. A dashed frame of the same color (blue or yellow or green or orange or purple) indicates strains isolated from the same patient. RAPD—random amplified polymorphic DNA, BAL—bronchoalveolar lavage, NBAL—non-bronchoscopic lavage (samples from lower respiratory tract other than bronchoalveolar lavage). In the case of the isolation of several *C. albicans* strains belonging to the same genotype from one patient (e.g., patient no. 28), only the strain isolated earliest is included in this Figure.

All *C. albicans* strains were sensitive to AMB, FCZ, VCZ, and ANF with MIC ranges of 0.062–0.25 mg/L, ≤0.125–0.5 mg/L, ≤0.008–0.031 mg/L, and ≤0.008–0.031 mg/L, respectively. Antifungal susceptibility profiles were similar in all genetic clusters (Table 3, Supplementary Information Table S1). Antifungal treatment was initiated in 10 out of 36 patients (28%) (Tables 1 and 2). In toral, 26 patients died (72%), including 7 (70%) who had received antifungal drugs. The fatality case rate for patients colonized with strains belonging to the largest clusters, i.e., no. 2, no. 1, no. 5, and no. 4, were 75%, 71%, 80%, and 75–100%, respectively (Table 2). The data for patient treatment obtained from the computer database were incomplete; for some cases, we do not have the information on the period of treatment, only the date of initiation of the therapy. In one of the two cases of invasive candidiasis confirmed by positive blood cultures (patient no. 35) for which there was complete treatment information available, treatment (3 days CAS, then 16 days FCZ) was started two days after sample collection and proved effective (the patient survived). In the second case (patient no. 28), the patient died despite the implementation of therapy (7 days CAS, then 5 days FCZ). Therapy was initiated 5 days after blood sample collection, which could be the reason for the lack of efficacy of the therapy, but there was a long period from the end of therapy to death (26 days), which raises doubts as to the cause of death. No further blood cultures were taken, but the same genotype found in the blood was isolated 2 and 7 days after the end of antifungal therapy from NBL samples (strains no. 98 and 101) (Tables 1 and 2). In the remaining cases in which antimycotics were used, *C. albicans* strains were isolated from LRT samples, and their clinical significance could not be determined retrospectively. In four cases (patients no. 5, 13, 20, and 25), antifungal therapy (VCZ or CAS or FCZ) was initiated 4 to 7 days after sample collection, and all cases ended in death. Additionally, patients no 5 and no 13 had co-infections with *C. glabrata*, which may also have contributed to death. In two patients (no. 12 and 18), FCZ was administered and the patients survived, but the database does not contain information on when therapy was initiated or how long it was administered. In patient no. 4, FCZ was administered two days after NBL sample collection and the patient died after 24 h of therapy, while in patient no. 22, FCZ was administered four days before NBL sample collection and the patient also died.

Some of the results were presented during the 57th Scientific Conference of the German speaking Mycological Society (DMykG), which took place in Frankfurt, Germany, on 27–29 September 2023.

Clusters no. 3 and 9–13 were represented by single *C. albicans* strains and are not included in the table. For these strains, the results of antifungal susceptibility testing are presented in Supplementary Information Table S1.

4. Discussion

Invasive candidiasis is a predominant fungal infection in hospitalized patients, in particular, among ICU patients, due to the accumulation of risk factors, and it is associated with high mortality [6]. Most candidiasis cases originate from the endogenous mycobiota of the patient's skin and mucous membranes; therefore, *Candida* strains can easily be transferred among hospital patients and personnel [22,28–33]. This could be one reason for the nosocomial outbreaks of invasive *Candida* infections. The period of the COVID-19 pandemic was especially challenging for the healthcare system. Due to the large number of patients requiring hospitalization and difficulties in access to basic protective equipment for medical staff, the horizontal transmission of microorganisms occurred with greater frequency [18–20,34–36]. In this study, two predominant clonal clusters of 17 *C. albicans* strains isolated from 12 patients and 12 *C. albicans* strains from 7 patients were identified with several smaller clusters, indicating the possibility of nosocomial transmission. Patients from whom *Candida* strains were isolated were hospitalized in one ward dedicated to COVID-19 patients over a period of 9 months. During this analysis, we were not able to determine exactly how *C. albicans* strains could translocate between patients. This is due to the nature of the study, which is retrospective. Genotyping was performed only for the strains that were isolated and deposited as part of routine hospital procedures, depending on the clinical condition of the patient, which did not include the testing of environmental samples or samples from medical personnel. During the period covered by this study, special sanitary rules were introduced in Poland, and it was not possible to conduct additional research. We also had limited clinical and workflow data, lacking information on patients' location, shared equipment, and shared medical staff. This information was not recorded in the database, so we do not have any additional evidence confirming the mode of transmission of the strains. Based on interviews with people working on the ward during that period, it appears that the severe clinical condition of patients sometimes required the use of sudden, unforeseen activities that could lead to the need to transfer patients around the ward and to accidental contacts promoting the spread and exchange of microbiota.

A major concern today in fungal healthcare-associated outbreaks is the spread of *C. auris* in the hospital environment, which have been reported worldwide [31–33,37–40]. We have not isolated *C. auris* strains although the incidence of COVID-19-associated candidiasis caused by this species is increasing in many hospitals, resulting in increased mortality among hospitalized patients with COVID-19 [15,41–47]. The first strains of *C. auris* were reported from Poland to ECDC in 2018 and 2019, but little is known about the real dissemination of this species in Polish hospitals [48]. Obviously, *C. auris* did not affect this hospital despite the COVID-19 emergence. Cases of the cross-transmission of other *Candida* species are not frequently reported but have been described before for *C. albicans*, *C. parapsilosis*, *C. tropicalis*, *C. lusitaniae*, *C. pelliculosa*, and *C. glabrata*, mostly occurring in neonatal and intensive care units [21–23,31–33,49,50]. Recently, apart from horizontal transmission and nosocomial infections of *C. auris*, emerging reports of fluconazole-resistant *C. parapsilosis* outbreaks have been described in, for example, Canada and Turkey [51,52]. Burnie et al. described a *C. albicans* outbreak of systemic candidiasis in 14 ICU patients in nine months [22]; 12 patients died. The strain responsible for this outbreak was also cultured from swabs from the oral mucous membrane and hands of medical staff, but it was not detected in the environment. This, therefore, indicates that healthcare personnel may play an important role in the transmission of *C. albicans* in hospitals.

Studying the genetic relationship between C. albicans and fungemia in the hospital can uncover the presence of endemic genotypes, which may suggest that the hands of healthcare personnel are important in the colonization and infection of *Candida* spp. [53]. Yildirim et al. (2007) found that overall, 34.1% of the hospital staff were found to harbor *Candida* spp. on their hands: 30.7% were nurses, 25.8% were resident doctors, 28.6% were laboratory workers, 84.6% were dining room personnel, and 43.3% were officers [54]. Thus, hospital workers, especially non-medical staff, should be educated in regular hand hygiene practice to prevent *Candida* colonization. Moreover, *Candida* can also colonize and persist in the hospital environment. *C. auris* was recovered from lanyards, temperature probes, floors, trollies, radiators, and windowsills. *C. parapsilosis* was found on neonatal incubators and humidifiers, patient beds, nurse workstations, computers, and floors, and *C. tropicalis* was found on haemodialysis machines. All these species have been reported to cause nosocomial epidemics [21]. In order to eliminate environmental reservoirs of *Candida*, attention should be paid to the quality of the disinfection of rooms, surfaces, and reusable materials, as well as the disinfectant used, its spectrum, and proper handling.

It would be of particular interest to repeat our study in the next period to check if the presence of the persistent endemic genotypes is confirmed and to follow their genetic evolution towards antifungal resistance. So far, all *C. albicans* strains tested in this study have been susceptible to antifungal drugs independently of where they originated. This finding may indicate that these strains do not belong to known worldwide clusters of resistant strains [55]. It also suggests that the selective pressure of antifungals used to treat invasive *C. albicans* infections has not been strong enough to lead to the presence of resistant strains with increased survival in the hospital environment. One possible factor might have been the limited use of antifungal drugs in the ICU (in this analysis, they were used in 28% of patients from whom *C. albicans* was isolated) and the short duration of therapy (in our study, in 40% of treated patients, therapy lasted for up to 3 days).

In Poland and the surrounding part of Europe, the research on fungal infections is still limited. Information on the prevalence of mycosis, species identification, patients' conditions during hospitalization and outpatient care, resistance, and genetics is sparse [56–60]. Fungal infections are only marginally treated, even in patients in the ICU or who are immunosuppressed. Increasing the awareness of fungal infections, as well as of the problem of horizontal transmission within hospitals, is crucial in decreasing the number of mycoses and fungal outbreaks. We hope that this study will draw more attention to fungal infections in our geographical area and will contribute to the more effective prevention, detection, and treatment of fungal infections.

5. Conclusions

Our study indicates that *C. albicans* can spread horizontally. The role of healthcare workers and the environment in the emergence of *C. albicans* nosocomial infections in critical areas through the unintentional spread of pathogens has not been confirmed because of limited data. The obtained results indicate the need to conduct a study covering not only hospitalized patients but also the environment and medical personnel. This will allow the confirmation of the horizontal transmission of *Candida* fungi, the determination of potential causes, and the introduction of preventive measures. It seems advisable to improve logistic and clinical data collection and banking, so that they can represent a reliable and complete source of information.

Supplementary Materials: The following supporting information can be downloaded at: https://www.mdpi.com/article/10.3390/jof10120864/s1, Figure S1: A-D: RAPD patterns of *Candida albicans* strains using OPA-18 primer. The corresponding strain numbers can be found in Table 2 with the genotypes. L correlates to the DNA standard (New England Biolabs). Table S1: Amphotericin B, fluconazole, voriconazole, and anidulafungin minimal inhibitory concentrations against *Candida albicans* strains belonging to different genetic clusters.

Author Contributions: All authors contributed to the study conception and design. M.S., M.N., A.S., M.G., J.J., J.Z. and J.W. collected, analyzed, and interpreted the retrospective data. M.S. performed antifungal drug susceptibility testing and interpreted the results. K.R., M.L. and R.W. genotyped the strains and interpreted the data. P.K. analyzed the data and prepared Figure 1. M.S. and J.W.-M. coordinated the project. M.S., K.R., P.B.H. and J.W.-M. prepared the manuscript draft. M.S. (corresponding author), K.R. and J.W.-M. reviewed and revised the manuscript and gave final approval to the version to be published. All authors have read and agreed to the published version of the manuscript.

Funding: This publication was supported by the National Centre for Research and Development through the initiative "Support for specialist hospitals in fighting the spread of SARS-CoV-2 infection and in treating COVID-19" (contract number: SZPITALE-JEDNOIMIENNE/18/2020) and by a grant from the Jagiellonian University Medical College, U1C/W41/NO/28.23. This study was carried out as part of our routine work. The research was supported in part by the doctoral program of excellence HOROS of the Austrian Science Fund (FWF, Vienna, Austria) grant ZFW12530 to the corresponding author R.W. For open access purposes, a CC BY public copyright license is used for any author-accepted manuscript version arising from this submission.

Institutional Review Board Statement: This work was approved by the Bioethics Committee of Jagiellonian University (approval no. 1072.6120.353.2020 from 16 December 2020). All the data analyzed during this study were anonymized prior to analysis.

Informed Consent Statement: The study was based on the laboratory data gathered during routine patients' care, and the analysis did not include any individual participant's data. As a result, no statements on consent from participants were required. The study in this form was approved by the local Bioethics Committee of Jagiellonian University.

Data Availability Statement: The datasets generated and/or analyzed during the current study are not publicly available but will be available from the corresponding authors on reasonable request.

Acknowledgments: We would like to thank the staff of the University Hospital in Krakow, especially the temporary Intensive Care Unit, for their help and interest in the study. Special thanks to Paweł Maga, Dept. of Angiology, University Hospital in Krakow, Poland.

Conflicts of Interest: The authors declare that they have no conflicts of interest.

References

1. Colombo, A.L.; Júnior, J.N.A.; Guinea, J. Emerging multidrug-resistant *Candida* species. *Curr. Opin. Infect. Dis.* **2017**, *30*, 528–538. [CrossRef]
2. Pappas, P.G.; Lionakis, M.S.; Arendrup, M.C.; Ostrosky-Zeichner, L.; Kullberg, B.J. Invasive candidiasis. *Nat. Rev. Dis. Primers* **2018**, *4*, 18026. [CrossRef]
3. Arendrup, M.C.; Patterson, T.F. Multidrug-resistant *Candida*: Epidemiology, molecular mechanisms, and treatment. *J. Infect. Dis.* **2017**, *216* (Suppl. 3), S445–S451. [CrossRef]
4. Barantsevich, N.; Barantsevich, E. Diagnosis and treatment of invasive candidiasis. *Antibiotics* **2022**, *11*, 718. [CrossRef] [PubMed]
5. Tharp, B.; Zheng, R.; Bryak, G.; Litvintseva, A.P.; Hayden, M.K.; Chowdhary, A.; Thangamani, S. Role of microbiota in the skin colonization of *Candida auris*. *mSphere* **2023**, *8*, e0062322. [CrossRef]
6. Soriano, A.; Honore, P.M.; Puerta-Alcalde, P.; Garcia-Vidal, C.; Pagotto, A.; Gonçalves-Bradley, D.C.; E Verweij, P. Invasive candidiasis: Current clinical challenges and unmet needs in adult populations. *J. Antimicrob. Chemother.* **2023**, *78*, 1569–1585. [CrossRef] [PubMed]
7. Logan, C.; Martin-Loeches, I.; Bicanic, T. Invasive candidiasis in critical care: Challenges and future directions. *Intensive Care Med.* **2020**, *46*, 2001–2014. [CrossRef]
8. Calandra, T.; Roberts, J.A.; Antonelli, M.; Bassetti, M.; Vincent, J.L. Diagnosis and management of invasive candidiasis in the ICU: An updated approach to an old enemy. *Crit. Care* **2016**, *20*, 125. [CrossRef]
9. Bassetti, M.; Giacobbe, D.R.; Vena, A.; Trucchi, C.; Ansaldi, F.; Antonelli, M.; Adamkova, V.; Alicino, C.; Almyroudi, M.-P.; Atchade, E.; et al. Incidence and outcome of invasive candidiasis in intensive care units (ICUs) in Europe: Results of the EUCANDICU project. *Crit. Care* **2019**, *23*, 219. [CrossRef]
10. Donnelly, J.P.; Chen, S.C.; Kauffman, C.A.; Steinbach, W.J.; Baddley, J.W.; Verweij, P.E.; Clancy, C.J.; Wingard, J.R.; Lockhart, S.R.; Groll, A.H.; et al. Revision and update of the consensus definitions of invasive fungal disease from the European Organization for Research and Treatment of Cancer and the Mycoses Study Group Education and Research Consortium. *Clin. Infect. Dis.* **2020**, *71*, 1367–1376. [CrossRef]
11. Bassetti, M.; Giacobbe, D.R.; Agvald-Ohman, C.; Akova, M.; Alastruey-Izquierdo, A.; Arikan-Akdagli, S.; Azoulay, E.; Blot, S.; Cornely, O.A.; Cuenca-Estrella, M.; et al. Invasive fungal diseases in adult patients in intensive care unit (FUNDICU): 2024 consensus definitions from ESGCIP, EFISG, ESICM, ECMM, MSGERC, ISAC, and ISHAM. *Intensive Care Med.* **2024**, *50*, 502–515. [CrossRef] [PubMed]
12. Bassetti, M.; Giacobbe, D.R.; Vena, A.; Wolff, M. Diagnosis and treatment of candidemia in the intensive care unit. *Semin. Respir. Crit. Care Med.* **2019**, *40*, 524–539. [CrossRef]
13. Bassetti, M.; Azoulay, E.; Kullberg, B.J.; Ruhnke, M.; Shoham, S.; Vazquez, J.; Giacobbe, D.R.; Calandra, T. EORTC/MSGERC Definitions of invasive fungal diseases: Summary of activities of the intensive care unit working group. *Clin. Infect. Dis.* **2021**, *72* (Suppl. 2), S121–S127. [CrossRef] [PubMed]
14. Negm, E.M.; Mohamed, M.S.; Rabie, R.A.; Fouad, W.S.; Beniamen, A.; Mosallem, A.; Tawfik, A.E.; Salama, H.M. Fungal infection profile in critically ill COVID-19 patients: A prospective study at a large teaching hospital in a middle-income country. *BMC Infect. Dis.* **2023**, *23*, 246. [CrossRef] [PubMed]
15. Hoenigl, M.; Seidel, D.; Sprute, R.; Cunha, C.; Oliverio, M.; Goldman, G.H.; Ibrahim, A.S.; Carvalho, A. COVID-19-associated fungal infections. *Nat. Microbiol.* **2022**, *7*, 1127–1140. [CrossRef] [PubMed]
16. Pemán, J.; Ruiz-Gaitán, A.; García-Vidal, C.; Salavert, M.; Ramírez, P.; Puchades, F.; García-Hita, M.; Alastruey-Izquierdo, A.; Quindós, G. Fungal co-infection in COVID-19 patients: Should we be concerned? *Rev. Iberoam. Micol.* **2020**, *37*, 41–46. [CrossRef] [PubMed]
17. Przyłęcki, P.; Wieczorkowska, M.; Pawlak-Kałuzińska, A.; Cedrowska-Adamus, W.; Gulczyńska, E. The COVID-19 pandemic impact on the Polish medical personnel work: A survey and in-depth interviews study. *Front. Public Health.* **2023**, *11*, 1187312. [CrossRef]
18. Arzilli, G.; Forni, S.; Righi, L.; Barnini, S.; Petricci, P.; Parisio, E.; Pistello, M.; Vivani, P.; Gemignani, G.; Baggiani, A.; et al. Trends in hospital acquired New Delhi metallo-beta-lactamase-producing *Enterobacterales* in Tuscany (Italy) from 2019 to 2021: Impact of the COVID-19 pandemic. *J. Hosp. Infect.* **2023**, *137*, 44–53. [CrossRef] [PubMed]
19. Monteiro, J.; Abboud, C.S.; Inoue, F.M.; Tufik, S.; Kiffer, C.R.V. NDM-producing *Enterobacterales* prevalence associated to COVID-19 in a tertiary hospital. *Infect. Dis.* **2023**, *27*, 102735. [CrossRef]
20. Patel, A.; Emerick, M.; Cabunoc, M.K.; Williams, M.H.; Preas, M.A.; Schrank, G.; Rabinowitz, R.; Luethy, P.; Johnson, J.K.; Leekha, S. Rapid spread and control of multidrug-resistant Gram-negative bacteria in COVID-19 patient care units. *Emerg. Infect. Dis.* **2021**, *27*, 1234–1237. [CrossRef] [PubMed]
21. Douglas, A.P.; Stewart, A.G.; Halliday, C.L.; Chen, S.C. Outbreaks of fungal infections in hospitals: Epidemiology, detection, and management. *J. Fungi.* **2023**, *9*, 1059. [CrossRef] [PubMed]
22. Burnie, J.P.; Odds, F.C.; Lee, W.; Webster, C.; Williams, J.D. Outbreak of systemic *Candida albicans* in intensive care unit caused by cross infection. *Br. Med. J. (Clin. Res. Ed.)* **1985**, *290*, 746–748. [CrossRef] [PubMed]
23. Lasheras, A.; Rogues, A.M.; Peyrere, S.; Boulard, G.; Bebear, C.; Gachie, J.; Bretagne, S.; Dromer, F. *Candida albicans* outbreak in a neurosurgical intensive care unit. *J. Hosp. Infect.* **2007**, *65*, 181–182. [CrossRef]

24. Lackner, M.; Najafzadeh, M.J.; Sun, J.; Lu, Q.; Hoog, G.S. Rapid identification of *Pseudallescheria* and *Scedosporium* strains by using rolling circle amplification. *Appl. Environ. Microbiol.* **2012**, *78*, 126–132. [CrossRef]
25. Bautista-Muñoz, C.; Boldo, X.M.; Villa-Tanaca, L.; Hernández-Rodríguez, C. Identification of *Candida* spp. by randomly amplified polymorphic DNA analysis and differentiation between *Candida albicans* and *Candida dubliniensis* by direct PCR methods. *J. Clin. Microbiol.* **2003**, *41*, 414–420. [CrossRef]
26. Arendrup, M.C.; Cuenca-Estrella, M.; Lass-Flörl, C.; Hope, W. EUCAST DEFINITIVE DOCUMENT E.DEF 7.3.2. Method for the determination of broth dilution minimum inhibitory concentrations of antifungal agents for yeasts EDef 7.2 (EUCAST-AFST). *Clin. Microbiol. Infect.* **2012**, *18*, E246–E247. [CrossRef] [PubMed]
27. European Committee on Antimicrobial Susceptibility Testing. Breakpoint Tables for Interpretation of MICs for Antifungal Agents. Version 10.0, Valid from 2020-02-04. Available online: https://www.eucast.org/fileadmin/src/media/PDFs/EUCAST_files/AFST/Clinical_breakpoints/AFST_BP_v10.0_200204_updatd_links_200924.pdf (accessed on 1 September 2023).
28. Pammi, M.; Eddama, O.; Weisman, L.E. Patient isolation measures for infants with candida colonization or infection for preventing or reducing transmission of *Candida* in neonatal units. *Cochrane Database Syst. Rev.* **2011**, *2011*, CD006068. [CrossRef]
29. Dorko, E.; Virágová, S.; Jautová, J.; Pilipčinec, E.; Danko, J.; Švicky, E.; Tkáčiková, L. Electrophoretic karyotyping of *Candida albicans* strains isolated from premature infants and hospital personnel in a neonatal intensive care unit. *Folia Microbiol.* **2001**, *46*, 453–457. [CrossRef] [PubMed]
30. Huang, Y.C.; Lin, T.Y.; Leu, H.S.; Peng, H.L.; Wu, J.H.; Chang, H.Y. Outbreak of *Candida parapsilosis* fungemia in neonatal intensive care units: Clinical implications and genotyping analysis. *Infection* **1999**, *27*, 97–102. [CrossRef]
31. Lupetti, A.; Tavanti, A.; Davini, P.; Ghelardi, E.; Corsini, V.; Merusi, I.; Boldrini, A.; Campa, M.; Senesi, S. Horizontal transmission of *Candida parapsilosis* candidemia in a neonatal intensive care unit. *J. Clin. Microbiol.* **2002**, *40*, 2363–2369. [CrossRef]
32. Finkelstein, R.; Reinhertz, G.; Hashman, N.; Merzbach, D. Outbreak of *Candida tropicalis* fungemia in a neonatal intensive care unit. *Infect. Control Hosp. Epidemiol.* **1993**, *14*, 587–590. [CrossRef]
33. Fowler, S.L.; Rhoton, B.; Springer, S.C.; Messer, S.A.; Hollis, R.J.; Pfaller, M.A. Evidence for person-to-person transmission of *Candida lusitaniae* in a neonatal intensive-care unit. *Infect. Control Hosp. Epidemiol.* **1998**, *19*, 343–345. [CrossRef]
34. Magnasco, L.; Mikulska, M.; Giacobbe, D.R.; Taramasso, L.; Vena, A.; Dentone, C.; Dettori, S.; Tutino, S.; Labate, L.; Di Pilato, V.; et al. Spread of carbapenem-resistant Gram-negatives and *Candida auris* during the COVID-19 pandemic in critically ill patients: One step back in antimicrobial stewardship? *Microorganisms* **2021**, *9*, 95. [CrossRef]
35. Alsuhaibani, M.; Kobayashi, T.; McPherson, C.; Holley, S.; Marra, A.R.; Trannel, A.; Dains, A.; Abosi, O.J.; Jenn, K.E.; Meacham, H.; et al. Impact of COVID-19 on an infection prevention and control program, Iowa 2020–2021. *Am. J. Infect. Control* **2022**, *50*, 277–282. [CrossRef]
36. Froidefond, M.; Sevestre, J.; Chaudet, H.; Ranque, S. COVID-19 is a confounder of increased *Candida* airway colonisation. *Pathogens* **2023**, *12*, 463. [CrossRef] [PubMed]
37. García, C.S.; Palop, N.T.; Bayona, J.V.M.; García, M.M.; Rodríguez, D.N.; Álvarez, M.B.; Serrano, M.D.R.G.; Cardona, C.G. *Candida auris*: Report of an outbreak. *Enferm. Infecc. Microbiol. Clin. (Engl. Ed.)* **2020**, *38* (Suppl. 1), 39–44. [CrossRef] [PubMed]
38. Du, H.; Bing, J.; Hu, T.; Ennis, C.L.; Nobile, C.J.; Huang, G. *Candida auris*: Epidemiology, biology, antifungal resistance, and virulence. *PLoS Pathog.* **2020**, *16*, e1008921. [CrossRef]
39. Biran, R.; Cohen, R.; Finn, T.; Tal, B.-N.; Galia, R.; Dafna, Y.; Sharon, A.; Yael, S.-M.; Alaa, A.; Jihad, B.; et al. Nationwide outbreak of *Candida auris* infections driven by COVID-19 Hospitalizations, Israel, 2021–2022. *Emerg. Infect. Dis.* **2023**, *29*, 1297–1301. [CrossRef] [PubMed]
40. Lyman, M.; Forsberg, K.; Sexton, D.J.; Chow, N.A.; Lockhart, S.R.; Jackson, B.R.; Chiller, T. Worsening spread of *Candida auris* in the United States, 2019 to 2021. *Ann. Intern. Med.* **2023**, *176*, 489–495. [CrossRef] [PubMed]
41. Bagheri Lankarani, K.; Akbari, M.; Tabrizi, R.; Vali, M.; Sekhavati, E.; Heydari, S.T.; Khodadadi, H.; Ahmadizar, F. *Candida auris*: Outbreak fungal pathogen in COVID-19 pandemic: A systematic review and meta-analysis. *Iran. J. Microbiol.* **2022**, *14*, 276–284. [CrossRef] [PubMed]
42. Vinayagamoorthy, K.; Pentapati, K.C.; Prakash, H. Prevalence, risk factors, treatment and outcome of multidrug resistance *Candida auris* infections in Coronavirus disease (COVID-19) patients: A systematic review. *Mycoses* **2022**, *65*, 613–624. [CrossRef]
43. Tsai, C.S.; Lee, S.S.; Chen, W.C.; Tseng, C.-H.; Lee, N.-Y.; Chen, P.-L.; Li, M.-C.; Syue, L.-S.; Lo, C.-L.; Ko, W.-C.; et al. COVID-19-associated candidiasis and the emerging concern of *Candida auris* infections. *J. Microbiol. Immunol. Infect.* **2023**, *56*, 672–679. [CrossRef] [PubMed]
44. Prestel, C.; Anderson, E.; Forsberg, K.; Lyman, M.; de Perio, M.A.; Kuhar, D.; Edwards, K.; Rivera, M.; Shugart, A.; Walters, M.; et al. *Candida auris* outbreak in a COVID-19 specialty care unit—Florida, July–August 2020. *MMWR Morb. Mortal. Wkly Rep.* **2021**, *70*, 56–57. [CrossRef]
45. Villanueva-Lozano, H.; Treviño-Rangel, R.J.; González, G.M.; Ramírez-Elizondo, M.T.; Lara-Medrano, R.; Aleman-Bocanegra, M.C.; Guajardo-Lara, C.E.; Gaona-Chávez, N.; Castilleja-Leal, F.; Torre-Amione, G.; et al. Outbreak of *Candida auris* infection in a COVID-19 hospital in Mexico. *Clin. Microbiol. Infect.* **2021**, *27*, 813–816. [CrossRef] [PubMed]
46. Chowdhary, A.; Tarai, B.; Singh, A.; Sharma, A. Multidrug-resistant *Candida auris* infections in critically ill Coronavirus disease patients, India, April–July 2020. *Emerg. Infect. Dis.* **2020**, *26*, 2694–2696. [CrossRef] [PubMed]
47. Chowdhary, A.; Sharma, A. The lurking scourge of multidrug resistant *Candida auris* in times of COVID-19 pandemic. *J. Glob. Antimicrob. Resist.* **2020**, *22*, 175–176. [CrossRef]

48. Plachouras, D.; Lötsch, F.; Kohlenberg, A.; Monnet, D.L. *Candida auris* survey collaborative group. *Candida auris*: Epidemiological situation, laboratory capacity and preparedness in the European Union and European Economic Area*, January 2018 to May 2019. *Euro Surveill.* **2020**, *25*, 2000240. [CrossRef]
49. Kalkanci, A.; Dizbay, M.; Turan, O.; Fidan, I.; Yalçin, B.; Hirfanoğlu, I.; Kuştimur, S.; Aktaş, F.; Sugita, T. Nosocomial transmission of *Candida pelliculosa* fungemia in a pediatric intensive care unit and review of the literature. *Turk. J. Pediatr.* **2010**, *52*, 42–49. [CrossRef] [PubMed]
50. Małek, M.; Paluchowska, P.; Bogusz, B.; Budak, A. Molecular characterization of *Candida* isolates from intensive care unit patients, Krakow, Poland. *Rev. Iberoam. Micol.* **2017**, *34*, 10–16. [CrossRef]
51. McTaggart, L.R.; Eshaghi, A.; Hota, S.; Poutanen, S.M.; Johnstone, J.; De Luca, D.G.; Bharat, A.; Patel, S.N.; Kus, J.V. First Canadian report of transmission of fluconazole-resistant *Candida parapsilosis* within two hospital networks confirmed by genomic analysis. *J. Clin. Microbiol.* **2024**, *62*, e0116123. [CrossRef] [PubMed]
52. Arastehfar, A.; Daneshnia, F.; Hilmioğlu-Polat, S.; Fang, W.; Yaşar, M.; Polat, F.; Metin, D.Y.; Rigole, P.; Coenye, T.; Ilkit, M.; et al. First report of candidemia clonal outbreak caused by emerging fluconazole-resistant *Candida parapsilosis* isolates harboring Y132F and/or Y132F+K143R in Turkey. *Antimicrob. Agents Chemother.* **2020**, *64*, e01001-20. [CrossRef] [PubMed]
53. Escribano, P.; Rodríguez-Créixems, M.; Sánchez-Carrillo, C.; Muñoz, P.; Bouza, E.; Guinea, J. Endemic genotypes of *Candida albicans* causing fungemia are frequent in the hospital. *J. Clin. Microbiol.* **2013**, *51*, 2118–2123. [CrossRef] [PubMed]
54. Yildirim, M.; Sahin, I.; Kucukbayrak, A.; Ozdemir, D.; Yavuz, M.T.; Oksuz, S.; Cakir, S. Hand carriage of *Candida* species and risk factors in hospital personnel. *Mycoses* **2007**, *50*, 189–192. [CrossRef]
55. Gong, J.; Chen, X.F.; Fan, X.; Xu, J.; Zhang, H.; Li, R.-Y.; Chen, S.C.-A.; Kong, F.; Zhang, S.; Sun, Z.-Y.; et al. Emergence of antifungal resistant subclades in the global predominant phylogenetic population of *Candida albicans*. *Microbiol. Spectr.* **2023**, *11*, e0380722. [CrossRef]
56. Mroczyńska, M.; Brillowska-Dabrowska, A. First report on echinocandin resistant Polish *Candida* isolates. *Acta. Biochim. Pol.* **2019**, *66*, 361–364. [CrossRef] [PubMed]
57. Nawrot, U.; Pajączkowska, M.; Fleischer, M.; Przondo-Mordarska, H.; Samet, A.; Piasecka-Pazik, D.; Komarnicka, J.; Sulik-Tyszka, B.; Swoboda-Kopeć, E.; Cieślik, J.; et al. *Candidaemia* in polish hospitals—A multicentre survey. *Mycoses* **2013**, *56*, 576–581. [CrossRef] [PubMed]
58. Deptuła, A.; Trejnowska, E.; Dubiel, G.; Wanke-Rytt, M.; Deptuła, M.; Hryniewicz, W. Healthcare associated bloodstream infections in Polish hospitals: Prevalence, epidemiology and microbiology-summary data from the ECDC Point Prevalence Survey of Healthcare Associated Infections 2012–2015. *Eur. J. Clin. Microbiol. Infect. Dis.* **2018**, *37*, 565–570. [CrossRef]
59. Gołaś, M.; Netsvyetayeva, I.; Sikora, M.; Piskorska, K.; Sulik-Tyszka, B.; Swoboda-Kopeć, E. Trends in antifungal susceptibility of *Candida* species--one year observation. *Pol. J. Microbiol.* **2014**, *63*, 217–222. [CrossRef] [PubMed]
60. Kurnatowska, A.; Kurnatowski, P.; Horwatt-Bożyczko, E.; Kurnatowska, A.J. Minimal inhibitory concentration (MIC) of caspofungin and itraconazole inhibiting growth of *Candida* strains calculated from the linear regression equation. *Adv. Med. Sci.* **2012**, *57*, 148–151. [CrossRef]

Journal of *Fungi*

MDPI

Review

Invasive Aspergillosis in the Intensive Care Unit

Anna Zubovskaia and Jose A. Vazquez *

Division of Infectious Diseases, Medical College of Georgia, Augusta University, Augusta, GA 30912, USA; azubovskaia@augusta.edu
* Correspondence: jvazquez@augusta.edu

Abstract: Invasive aspergillosis (IA) is a fungal infection, which has traditionally been associated with neutropenia and immunosuppressive therapies. Our understanding of invasive aspergillosis has been evolving and, in the past few decades, IA among ICU patients has been recognized as a common infection and has become more widely recognized. The diagnosis and management of invasive aspergillosis in the ICU is particularly challenging, due to the unstable clinical condition of the patients, lack of diagnostic markers, increased risk of further clinical deterioration, multiple comorbidities, and a need for early assessment and treatment. In this article, we will discuss the challenges and pitfalls of the diagnosis and management of invasive aspergillosis in an ICU setting, along with a review of the current literature that is pertinent and specific to this population.

Keywords: aspergillosis; intensive care unit; invasive aspergillosis; antifungals

1. Introduction

Invasive aspergillosis (IA) is an infection caused by a saprophytic filamentous fungus of the *Aspergillus* species [1]. It is a known opportunistic infection in immunocompromised hosts, such as patients with acute myelogenous leukemia (AML), prolonged neutropenia, solid organ transplant (SOT) recipients and allo-stem cell transplant recipients (allo-SCT) [2]. However, invasive aspergillosis is an increasingly recognized infection among patients in the intensive care unit, often developing in patients without classic risk factors and complicating other respiratory diseases such as COPD, severe COVID-19 and influenza [1–3]. Patients with COVID-19-associated pulmonary aspergillosis (CAPA) and influenza-associated pulmonary aspergillosis (IAPA) have a higher mortality than patients without CAPA/IAPA [3–5]. In fact, when compared to other respiratory tract infections, the mortality among patients in the ICU who developed IA can be as high as 46% [6]. The diagnosis of invasive aspergillosis and critically ill patients remains challenging. In this review, we will discuss the epidemiology of IA among ICU patients and the current evidence-based approach to the diagnosis and management of IA in the ICU, including IAPA and CAPA.

2. Epidemiology

The diagnostic challenges and the low suspicion among providers make IA a difficult diagnosis to establish and thus, make the true incidence of IA among ICU patients difficult. However, the prevalence of IA among severely ill patients and ICU patients seems to be increasing including patients with COVID-19, influenza virus and COPD. Solid organ transplant status, solid organ tumors and prolonged exposure to corticosteroids, as well as the use of ibrutinib, are increasingly recognized as risk factors for invasive aspergillosis.

The incidence of invasive aspergillosis in the intensive care unit has been reported to be between 0.33% and 7% [1,2,5–9]. Reviewing the service level database including inpatient data from the years 2005–2008, 6.4% of ICU patients were diagnosed with IA [6]. In this study, evaluating the diagnostic accuracy of the Asp-ICU algorithm, 71% of patients with proven invasive aspergillosis were immunocompromised and 70% of patients with putative IAPA had known risk factors [10]. *Aspergillus fumigatus* is the most commonly encountered species among the ICU patients, isolated in 92% of the cases, followed by *A. flavus*, *A. niger* and other species [11]. *A. fumigatus* was also the most frequently encountered species among patients with IAPA, as well as CAPA [12,13]. In this study, the predictive score for the development of CAPA was evaluated and *A. fumigatus* was isolated in 64.3% of cases, followed by *A. niger* complex, less frequently *A. terreus* and *A. flavus* [13].

Influenza-associated pulmonary aspergillosis (IAPA) is a complication of severe influenza which causes increased morbidity and mortality compared to patients without IAPA. The incidence of IA ranges between 10 and 32% of influenza patients admitted to the ICU [14–16].

Invasive aspergillosis-complicating COVID-19 infection (CAPA) is an increasingly recognized phenomenon. Based on the published data, the prevalence of CAPA has been estimated to be between 5 and 30% [17].

Increased mortality among patients with invasive aspergillosis in the ICU, IAPA and CAPA has been described in multiple studies [2,16–18].

3. Risk Factors and Pitfalls in IA Definitions in the ICU

Known and classic risk factors of IA include immunosuppression, such as prolonged neutropenia due to leukemia or other hematological malignancy, because of chemotherapeutic agents or as a result of HSCT, as well as acute GVHD grade 3 or 4, involving either the gastrointestinal tract, lungs or liver that is refractory to first-line treatments such as steroids. The duration and severity of neutropenia are also associated with an increased risk of IPA [19,20]. Solid organ transplant recipients (specifically, lung and heart transplant recipients) and the prolonged use of corticosteroids (equal to or above 0.3 mg/kg corticosteroids for equal to or above 3 weeks in the past 60 days) are other well-known predisposing conditions [1,21]. Additional risk factors for the development of IA include treatment with T-cell immunosuppressants, such as calcineurin inhibitors, TNF alpha blockers, lymphocyte-specific monoclonal antibodies, or immunosuppressive nucleoside analogs for 90 days. In addition, treatment with a recognized B-cell immunosuppressant, such as Bruton's tyrosine kinase inhibitors (ibrutinib), has also been associated with IA. Inherited severe immunodeficiencies, such as chronic granulomatous disease, STAT3 deficiency or SCID (severe combined immunodeficiency), can become predisposed to invasive pulmonary mold disease.

These are uniformly recognized host risk factors included in the consensus definition of invasive fungal disease published by EORTC/MSG [21]. However, patients in the ICU setting lack traditional risk factors and different pathogenic mechanisms predispose them to invasive aspergillosis in the ICU setting (Table 1) [11]. This poses a significant diagnostic challenge since the ICU patients who do not meet the EORTC/MSG criteria cannot fulfill the criteria used to diagnose invasive aspergillosis. Furthermore, tissue sampling in these settings may be difficult or contraindicated in patients in the ICU with hemodynamic instability, coagulopathy or thrombocytopenia. Moreover, the classic radiographic signs of IA, such as the halo sign and air crescent sign, are infrequently encountered in patients without classic risk factors (most ICU patients) and *Aspergillus* serum galactomannan testing in this population has been shown to have low sensitivity and specificity. In addition, the low importance of invasive aspergillosis in the ICU setting, represents another diagnostic challenge [22].

Table 1. Risk factors for invasive aspergillosis in the ICU [7].

High Risk	Intermediate Risk	Low Risk
Neutropenia (neutrophil count <500/mm^3)	Prolonged treatment with corticosteroids before admission to the ICU	Severe burns
Hematologic malignancy	Autologous bone marrow transplantation	Other solid organ transplant recipients (e.g., heart, kidney or liver)
Allogeneic bone marrow transplantation	COPD	Steroid treatment with a duration of ≤7 days
	Liver cirrhosis with a duration of stay in the ICU >7 days	Prolonged stay in the ICU (>21 days)
	Solid organ cancer	Malnutrition
	HIV infection	Post-cardiac surgery status
	Lung transplantation	
	Systemic diseases requiring immunosuppressive therapy	

The Asp-ICU score was developed to capture cases of IAPA in non-neutropenic ICU patients who could not fulfill the criteria classic EORTC/MSG criteria [10]. Autopsy results among deceased ICU patients indicated that adherence to host factors for invasive fungal disease as one of the key features used to meet the criteria for invasive fungal disease increased the risk of a missed diagnosis [8]. High mortality rates among patients with CAPA/IAPA in the ICU have been found in the absence of traditional risk factors, such as neutropenia, SCT or other types of immune deficiency. The ASP ICU diagnostic algorithm was created based on EORTC/MSG criteria for invasive fungal disease; however, the interpretation of imaging and microbiological data was modified to increase its utility in diagnosing IA specifically among ICU patients [8]. The entry criterion was positive lower respiratory culture, and the probability of a true invasive infection was assessed by evaluating the clinical and radiological data in conjunction with the positive culture. Clinical criteria, such as recrudescent fever despite at least three days of appropriate antimicrobial therapy and no other apparent source, as well as worsening respiratory insufficiency despite appropriate antibiotic therapy and ventilatory support, were specifically tailored to ICU patients. The radiological criterion (abnormal X-ray or CT findings) appreciated the frequently nonspecific imaging findings among ICU patients with IA and low sensitivity of the classic radiological signs, such as the air crescent sign or halo sign in the ICU population, and increased the sensitivity of the algorithm among the ICU cohort. Finally, the presence of the classic host factors, traditionally considered a prerequisite for the diagnosis of IA, although still present among the criteria, was no longer a requirement [8]. The algorithm demonstrated a higher sensitivity and negative predictive value in patients without classic risk factors as well as in immunocompromised hosts [10].

The original ASP ICU algorithm proposed over a decade ago did not include fungal biomarkers, such as galactomannan and the *Aspergillus* quantitative PCR. A revision of the algorithm was later proposed to improve the detection of probable invasive pulmonary aspergillosis in the ICU (Table 2) [23]. These revisions are now included in the mycological criteria and the definitions of invasive pulmonary aspergillosis in ICU patients, including IAPA, CAPA and invasive fungal disease from EORTC/MSG (Table 2) [21,23].

Table 2. Clinical and radiological features of probable invasive aspergillosis in the ICU setting [7,12,21,22].

Pulmonary aspergillosis
The presence of one of the following patterns on a CT:
• Dense, well-circumscribed lesions with or without a halo sign (typical in neutropenic patients). The halo sign is often not applicable in the ICU as the sign arrives too early (5 days before the onset of disease); not specific for *Aspergillus*
• Air crescent sign—often not applicable in the ICU because it is obscured by atelectasis, ARDS and/or pleural effusion
• Cavity
• Wedge-shaped and segmental or lobar consolidation
• Bronchopneumonia
• Nodular disease
Tracheobronchitis
• Tracheobronchial ulceration, pseudomembranes, nodule, plaque or eschar detected by bronchoscopy
Sino-nasal diseases
• Acute localized pain (including pain radiating to the eye)
• Nasal ulcer with black eschar
• Extension from the paranasal sinus across bony barriers, including into the orbit
Central nervous system infection
One of the following two signs:
• Focal lesions on imaging
• Meningeal enhancement on MRI or CT

The current diagnostic classification of invasive fungal disease according to the scale of certainty still requires the positive culture of the *Aspergillus* species along with a histopathological examination of the tissue obtained in a sterile fashion from a normally sterile site and with evidence of accompanying tissue damage. However, as previously discussed, such invasive methods are not always appropriate in the ICU setting. Thus, the diagnosis of probable invasive aspergillosis appears to be a more applicable criterion in this type of setting (Table 3).

Considering the challenges in diagnosing and defining cases of IA in an ICU setting, as well as the varying prevalence of IA across hospitals and the varying frequency of proven diagnoses, revised definitions of proven and probable invasive pulmonary aspergillosis in the ICU setting were proposed [22]. Revised definitions incorporated mycological evidence, clinical/radiological abnormality (see Table 2, sections regarding pulmonary aspergillosis and tracheobronchitis) and host factors. Mycological evidence for proven and probable IA was proposed. Additional host factors have been proposed to establish the diagnosis of IA in the ICU population. These risk factors reflect the unique pathogenesis of IA in the ICU population, in addition to the classic risk factors, such as SOT, neutropenia, hematological malignancies and HSCT. These additional risk factors include chronic respiratory airway abnormalities such as COPD or bronchiectasis; decompensated cirrhosis; HIV; treatment with recognized immunosuppressants (e.g., calcineurin or mammalian target of rapamycin [mTOR] inhibitors, blockers of tumor necrosis factor [TNF] and similar antifungal immunity pathways, alemtuzumab, ibrutinib, or nucleoside analogs) during the previous 90 days; glucocorticoid treatment with prednisone equivalent of 20 mg or more per day; and severe viral pneumonias such as influenza and COVID-19 [22]. Additionally, patients with severe burns, a prolonged ICU stay, and a positive fungal culture are also at risk for developing IA in the ICU [24].

Table 3. Mycological evidence supporting the diagnosis of probable invasive aspergillosis (based on updated EORTC/MSG criteria [21]).

Invasive pulmonary Aspergillosis
• *Aspergillus* recovered by culture from sputum, BAL, bronchial brush or aspirate
• Microscopical detection of fungal elements in sputum, BAL, bronchial brush or aspirate indicating *Aspergillus*
Tracheobronchitis
• *Aspergillus* recovered by the culture of BAL or bronchial brush
• Microscopic detection of fungal elements in BAL or bronchial brush indicating *Aspergillus*
Sino-nasal diseases
• *Aspergillus* recovered by the culture of sinus aspirate samples
• Microscopic detection of fungal elements in sinus aspirate samples indicating *Aspergillus*
Galactomannan antigen
Antigen detection plasma, serum, BAL or CSF
• Any one of the following:
• Single serum or plasma: ≥1.0
• BAL fluid: ≥1.0
• Single serum or plasma: ≥0.7 and BAL fluid ≥0.8
• CSF: ≥1.0
Aspergillus PCR
• Any one of the following:
○ Plasma, serum or whole blood from two or more consecutive PCR tests is positive
○ BAL fluid from two or more duplicate PCR tests positive
○ At least one positive PCR test in plasma, serum or whole blood and one positive PCR test in BAL fluid

Invasive aspergillosis-complicating viral pneumonia, such as influenza (IAPA) or COVID-19 (CAPA) is not a rare infection encountered among ICU patients. According to the study by Waldeck et al., 10.8% of 158 patients hospitalized in tertiary hospitals in Switzerland during the 2017/2018 and 2019/2020 influenza seasons who were admitted with a PCR were confirmed to have an influenza infection, required ICU admission for over 24 h and developed IAPA. In addition, those with a prior history of asthma and days of mechanical ventilation were also associated with the development of IAPA [16]. Several other studies show varying rates of IAPA up to 28.1%. Influenza was found to be an independent risk factor in the development of IA, conferring a high mortality among ICU patients [12,15,25,26].

Aspergillus hyphae are present in the airways and frequently represent colonization [20]. The key feature that determines the pathogenesis of invasive aspergillosis is the transition from colonization to invasion of the underlying tissues, specifically angioinvasion [27]. In the susceptible host, such as patients in the ICU or those with a viral infection of the upper respiratory tract, increased production of inflammatory cytokines leads to local tissue damage, which aids in further tissue invasion by *Aspergillus*. Several research groups suggested that respiratory viruses, such as SARS-CoV-2 and influenza, cause lung hyperinflammation, which causes defects in several levels of antifungal immunity, including the phagocytosis of *Aspergillus* conidia, the epithelial barrier and function and the neutrophil-mediated killing of *Aspergillus* hyphae [28]. In COVID-19, a decrease in T-cell populations, especially in patients with severe disease, has been observed, probably accompanied by defective lymphocyte function, increasing the risk of invasive mold infection [29,30]. Angioinvasion occurs in the pulmonary vasculature, as well as in other organs during the disseminated infection, and results in thrombosis and tissue infarction, as well as reduced leucocyte entry into the infected area and reduced delivery of antifungals [27].

There is a significant variation in estimates of the incidence of COVID-19-associated pulmonary aspergillosis (CAPA). According to the study by Hurt et al. conducted across five UK hospital intensive care units among patients admitted for mechanical ventilation or ECMO due to respiratory failure from COVID-19, the incidence of probable CAPA was 10.9%, followed by 5.2% of the study population with possible CAPA. No definitive cases of CAPA were found in this study [31]. In a multinational observational study conducted by the European Confederation of Medical Mycology (ECMM), the median prevalence of CAPA per enrolled center was 10.7% (range 1.7–26.8%) [18]. However, in the review of the autopsy series, which included case studies describing autopsies from 677 subjects, of which 320 were on mechanical ventilation, only 11/677 (2%) had an invasive mold infection. This included eight cases of CAPA, an unspecified number of invasive mold diseases and one disseminated mucormycosis. Among those who received mechanical ventilation (320/677), only 6 (2%) had an invasive mold infection [32].

In a retrospective cohort study from Johns Hopkins, the author's goals were to design a CAPA prediction model using mechanically ventilated patients to stratify which patients were at risk of developing CAPA and who would benefit from additional testing and antifungal treatment. Age, time from intubation, use of dexamethasone for COVID-19 treatment, underlying pulmonary circulatory disease, multiple myeloma, cancer or hematologic malignancy were identified as risk factors for CAPA and were included in the prediction model. In that study, 11.8% (98 patients) of the cohort met criteria for CAPA. Patients with CAPA in this cohort had a higher mortality or were more likely to require advanced life support and to have a longer duration of advanced life support therapy [17].

Another prediction model was developed based on the retrospective matched case–control study conducted at a tertiary care center in South India [33]. The European Organization for Research and Treatment of Cancer Risk Factors used this prediction score to identify lymphopenia and broad-spectrum antimicrobials as the main risk factors for CAPA. Of note, due to low enrollment numbers, the study failed to identify hematologic malignancies, SOT and T-cell and B-cell immunosuppressants, as well as the use of ibrutinib, as risk factors for the development of CAPA. The recovery of bacterial pathogens from blood and respiratory secretions frequently decreased the risk of CAPA. All cases in the study were either probable or possible CAPA. In addition, the study included both mechanically ventilated and nonventilated patients. The diagnosis of CAPA increased the mortality rate and ICU admission, however, it did not significantly increase the risk of requiring mechanical ventilation. The isolation of bacterial pathogens in blood or BAL decreased the likelihood of CAPA. The sensitivity of this model was 77.4% with a specificity of 78.1%. Despite certain limitations, this model might hold promise in low-resource settings [33].

Another study published in 2024 by Iacovelli et al. evaluated risk factors for the development of CAPA in a respiratory sub-intensive care unit and its impact on overall mortality. Hematological malignancy, lymphocytopenia and COPD were identified as independent risk factors for CAPA. Furthermore, being over 65 years was identified as a predictor of mortality [34].

Another CAPA clinical prediction score based on a study by Calderón-Parra et al. from Spain, identified old age, active smoking, chronic respiratory disease, chronic renal disease, chronic corticosteroid treatment, tocilizumab therapy and a high Apache 2 score on admission as risk factors for CAPA [13].

In the recently published meta-analysis by Gioia et al., nine risk factors for CAPA were identified including chronic liver disease, neurological malignancy, chronic obstructive pulmonary disease, cerebrovascular disease and diabetes, as well as mechanical ventilation, the use of renal replacement therapy, the treatment of COVID-19 with interleukin-6 inhibitors and the treatment of COVID-19 with corticosteroids. Patients with CAPA were typically older and the duration of mechanical ventilation was longer among patients with CAPA than those without CAPA [35].

4. Clinical Presentation

The clinical presentation of invasive aspergillosis in the ICU setting frequently overlaps underlying co-existing comorbidities, such as COVID-19, influenza, pneumonia, sepsis, etc., which poses an additional diagnostic challenge [5,20]. Fever, cough, hemoptysis and a pleuritic chest can be the presenting symptoms of invasive pulmonary aspergillosis; however, these symptoms are nonspecific and might not be present in all cases [1,7,20]. The development of sinonasal or CNS aspergillosis might also be accompanied by the signs and symptoms of local or regional spread of the infection. There are significant differences in the clinical presentation between neutropenic and non-neutropenic patients [36]. Specifically, IA in non-neutropenic patients is generally less symptomatic and frequently associated with pneumonia due to another organism (viral) and associated with higher mortality rates [36]. Moreover, IA in non-neutropenic patients who had high serum GM levels were more likely to have COPD and had more severe respiratory symptoms, which included hemoptysis and dyspnea [37].

5. Diagnosis of Invasive Aspergillosis in the ICU

Histopathological diagnosis is based on the presence of tissue invasion and *Aspergillus* growth from the normal sterile site (Figure 1a,b) [38].

However, despite being the gold standard, positive *Aspergillus* cultures used as a diagnostic method have a very low diagnostic yield. This is even lower among non-neutropenic ICU patients who lack traditional host risk factors. In addition, microscopy and culture alone cannot distinguish between colonization and infection [39,40]. As previously discussed, biopsies obtained from sterile sites are required to fulfill the histopathological criteria of proven invasive aspergillosis. These procedures may be not feasible among some ICU patients due to unstable clinical conditions and contraindications to invasive procedures [2,22]. Due to these limitations, a less invasive approach to establishing a diagnosis is usually utilized [22]. Frequently, a diagnosis is presumptive and made based on imaging findings, serum biomarkers, sputum and/or BAL specimens and utilizes nonculture-based methods [22]. The sensitivity of BAL cultures differs between neutropenic and non-neutropenic patients [36].

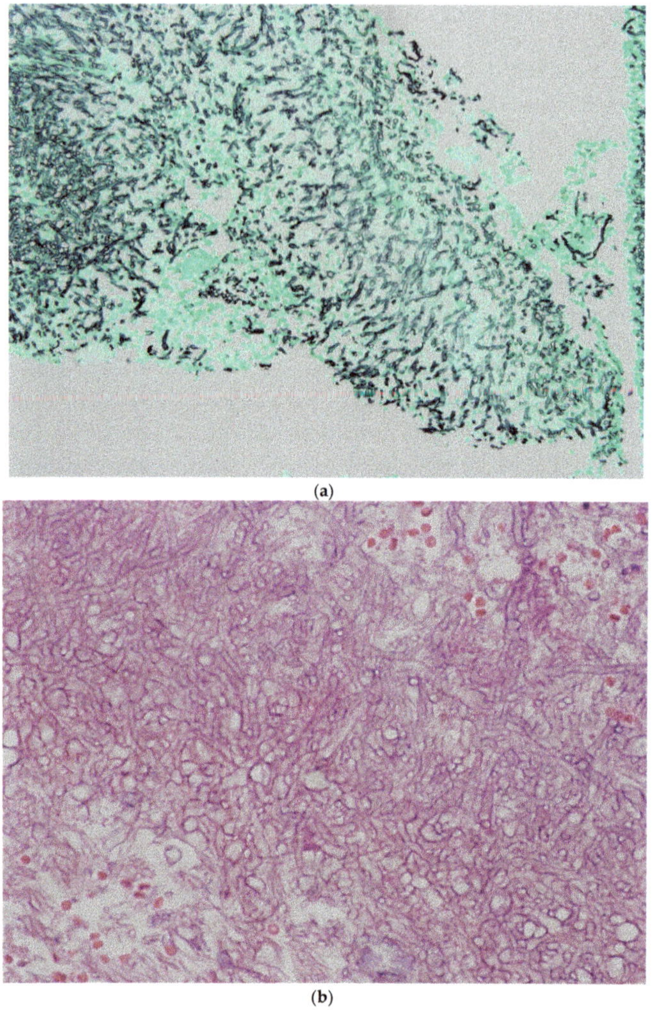

Figure 1. (**a**) Aspergillus in tissue, Grocott Methenamine stain (×250). (**b**) Aspergillus in tissue, H&E stain (×400).

Galactomannan is an *Aspergillus*-specific antigen, a major component of the *Aspergillus* cell wall. It can be measured in serum, plasma, BAL samples or CSF. Studies investigating the utility of galactomannan testing on upper airway samples have been previously published [38,39]. During the COVID-19 pandemic, alternative samples from the upper respiratory tract, such as non-bronchoalveolar lavage, tracheal aspirate and sputum, were suggested as an alternative to the more invasive BAL samples for obtaining Aspergillus cultures and galactomannan due to the restrictions imposed by the COVID-19 pandemic. However, these biomarkers have not been validated in tracheal aspirates, sputum and non-bronchoalveolar lavages. Since cut-off values have not been well established, results have to be interpreted with caution [41–43]. The angioinvasion of *Aspergillus* can result in galactomannan being released into the bloodstream and becomes detectable in serum or plasma. However, it is often not found in the serum of non-neutropenic patients where more airway invasions rather than angioinvasion are present [43,44]. The performance of the serum GM is also negatively affected by concurrent systemic anti-mold therapy, which may yield false-negative results [45]. False-positive results of the serum galactomannan assay have been described in patients who received certain antimicrobials, such as amoxicillin/clavulanate, piperacillin tazobactam and cefepime. False-positive results of galactomannan in the BAL have also been reported in patients receiving carbapenems and ceftriaxone [46]. Among patients post-HSCT within the last 100 days and in those with GI GVHD, false-positive results can be encountered due to compromise of the mucosal barrier and the translocation of the galactomannan through the intestinal mucosa [47]. Other fungal species (*Penicillium* spp., *H. encapsulatum*, and *Geotrichum* spp., etc.), containing galactomannan in their cell walls can also yield false positive results [48]. Among non-neutropenic patients, a positive GM in serum is encountered less often and has lower sensitivity, most likely due to less prominent angioinvasive features in these cases [29,37].

Aspergillus PCR in blood, serum or BAL were included in the mycological criteria and may be a useful tool to establish the diagnosis of invasive aspergillosis. According to the 2019 meta-analysis published by Cruciani et al., pooled sensitivity and specificity of PCR from blood is reported to be between 79% and 80% for a single positive result and 60% and 94% for two consecutive positive test results in immunocompromised people [49]. The performance of PCR in blood decreases in the setting of systemic mold-active therapy, as well as in non-neutropenic patients [50,51].

Aspergillus PCR in BAL is a promising tool for diagnosing IA in neutropenic and non-neutropenic patients, and it is generally not affected by antifungal therapy [52]. Various assays demonstrate varying degrees of sensitivity in immunocompromised patients [53,54]. However, it appears to have excellent specificity in critically ill patients with and without COVID-19 and/or immunocompromising conditions; although, the study by Mikulska et al. showed that sensitivity is higher among immunocompromised patients [54]. The sensitivity and specificity improve when several diagnostic tests are used in combination, including BAL-GM, BAL-PCR, serum GM and BAL cultures [55].

In recent years, the utility of metagenomic next-generation sequencing (NGS) and cell-free DNA is being actively explored, both in patients with hematological disorders and in those without immunocompromising conditions [56]. Next-generation sequencing of microbial cell-free DNA using the Karius test on plasma as compared to the standard of care procedures revealed a sensitivity of 38.5% and a specificity of about 97% in patients with probable IA. In addition, the results are affected by antimould therapy [57]. *Aspergillus* plasma cell-free DNA PCR showed superiority over *Aspergillus* serum galactomannan in the diagnosis of invasive aspergillosis among patients with hematological illnesses/stem cell transplants, demonstrating an overall sensitivity of 86% and a specificity of 93.1% (Table 4) [58].

The utility of metagenomic next-generation sequencing for the diagnosis in non-neutropenic patients at risk of invasive aspergillosis, including CAPA, is being explored [59–63]. The utility of blood biomarkers such as galactomannan is often poor in CAPA, as well as invasive pulmonary aspergillosis in non-neutropenic patients, as it exhibits early tissue-invasive growth in the lungs with delayed angioinvasion. The Karius test showed promising performance, with a high specificity of 97% [60].

More recently, the lateral flow assay (LFA) and the lateral flow device (LFD) have become point-of-care diagnostic tests for the diagnosis of invasive aspergillosis. They show good performances on both serum samples and BALF and aid in prompt diagnosis of IA.

GM-LFA shows excellent performance in patients with hematological malignancies [64]. The performance in SOT recipients can be variable [65]. Several LFDs are being studied for clinical use among neutropenic patients, as well as SOT recipients and patients in the ICU [66–68].

Table 4. Utility of mcfDNA (microbial cell-free DNA) analysis and NGS in diagnosis of invasive pulmonary aspergillosis.

Study, First Author	# of Patients	Population	Tested Sample	Sensitivity	Specificity	PPV	NPV	Additional Data
Liu, 2024 [61]	N = 66, 21 with IPA, 45 non-pulmonary aspergillosis	Patients with T2DM, with and without immuno-compromising conditions	BALF (90.5%), blood (9.5%)	66.7%	100%	100%	86.5%	Significantly improved performance when combined with other methods
Bao, 2022 [62]	N = 33, 12 with IPA, 21 non-pulmonary aspergillosis	Non-neutropenic patients	BALF: N = 27 Blood: N = 6 Pleural fluid N = 1 BALF+serum: N = 3	91.7%	71.4%	64.7%	93.8%	Additionally evaluated co-pathogens in mixed infections, performance reported as cumulative for mNGS on all types of samples
Huygens, 2024 [57]	N = 106, proven/probable IA: N = 35, IA+other IFD: N = 4, other IFD = 7, possible IFD = 48	AML, MDS, HSCT, hematological malignancies, neutropenia	Karius test—plasma: N = 106, research-only pipeline Karius on BALF, N = 34	Plasma: 44% RUO-BAL KT: 72.2%	Plasma:96.6% RUO-BAL KT: 88.2	NA	NA	Data for multiple IFD, performance not impacted by mold active therapy
Lee, 2024 [63]	N = 34, N = 1 (proven IA), N = 25 (probable), N = 3 (putative), N = 5 (no IA)	Hematological malignancies = 16, COVID-19 = 19.	Plasma cfDNA, N = 34	N/A	N/A	N/A	N/A	Concordance between cfDNA and conventional methods of diagnosis is higher in HM group than in COVID-19
Hoenigl, 2023 [60]	N = 114, CAPA: proven: N = 0, probable: N = 6, possible: 2, No CAPA: N = 106	COVID-19 associated ARF in the ICU	Karius, plasma mcfDNA	67% for Aspergillus, 83% for other molds	97% for patients without CAPA	N/A	N/A	
Mah, 2023 [58]	N = 238, Proven: N = 15, probable: N = 31, possible: N = 62, no IA: N = 130	89.9% immuno-suppressed (HM, HSCT, SOT)	Plasma cfDNA	Overall, 80.0% (varies in different groups)	Overall, 93.1%	5% prevalence: 39.6% 20% prevalence: 75.7%	5% prevalence: 99.2% 20% prevalence: 96.4%	Highest performance in HM/HSCT patients, performance characteristics vary by patient groups

Abbreviations: AML—acute myelogenous leukemia; ARF—acute respiratory failure; BALF—bronchoalveolar lavage fluid; CAPA—COVID-19-associated pulmonary aspergillosis; HM—hematological malignancy microbial cell-free DNA; mNGS—metagenomic next-generation sequencing; RUO-BAL KT—research-use-only Karius test on bronchoalveolar lavage fluid; SOT—solid organ transplant.

The utility of beta-D-glucan in the diagnosis of invasive aspergillosis remains unclear since it is a cell wall component of many fungi and it is not specific for *Aspergillus* species. Blood levels are often elevated in the setting of gut fungal translocation due to compromised intestinal epithelium due to diarrhea, GI Tract immune dysfunction or altered gut microbiota [69–71]. False-positive results can occur in those receiving albumin, and for hemodialysis with [illegible] in those receiving IVIG and IV amoxicillin–clavulanic acid [72].

6. Imaging

The radiographic features of invasive pulmonary aspergillosis are variable and frequently nonspecific. The typical radiologic manifestations of IFD include nodules, masses, segmental or subsegmental consolidations, atelectasis, ground-glass opacities, a tree-in-bud pattern, cavities or pleural effusions [73]. Invasive pulmonary aspergillosis may appear as solitary or multiple pulmonary nodules/masses on chest X-rays or possibly wedge-like areas of ill-defined opacities. Chest X-rays have poor sensitivity and lack the ability to differentiate IPA from other etiologies of pneumonia, so a CT chest scan is the imaging modality of choice [74]. The classical findings in CT chest scans include multiple nodules with a "halo" sign, which is an area of ground-glass opacity surrounding a pulmonary nodule [74]. The ground-glass component represents angioinvasion and hemorrhage into the area surrounding the fungal nodule. An area of cavitation may develop in the nodule, which is classically described as an "air crescent sign" (Figure 2a,b).

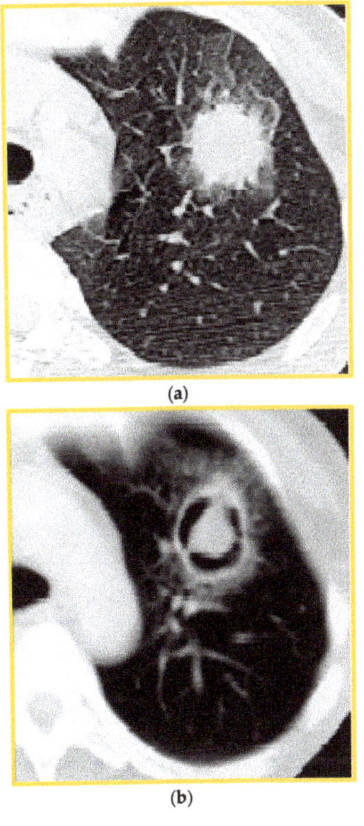

Figure 2. (**a**) Invasive pulmonary aspergillosis on HRCT. "Halo" sign. (**b**) Invasive pulmonary aspergillosis on HRCT. "Air crescent sign". (HRCT = High-resolution Cat Scan).

Although both findings have been described as classical in invasive pulmonary aspergillosis, the most common presentation among immunocompromised patients is one or more macronodules in 94% of patients, while the halo sign is only present in approximately 61% of patients, with consolidations present in 30%, cavitary lesions in 20% of patients and, less commonly, with an air crescent sign (10%) [75]. The air crescent sign is a rare finding in early pulmonary aspergillosis, and it may not assist in the early diagnosis of IAPA. In neutropenic patients, it can be detected after the partial recovery of the neutrophil function [76]. These two classic findings described in immunocompromised patients are far less frequently encountered in non-neutropenic patients. Among 116 patients with non-neutropenic IPA, the most common findings included consolidation (47.4%) and cavities (47.4%), whereas the air crescent sign was detected in only 14.7% and the halo sign in 3.4% [37].

Despite being the imaging modality of choice, aiding in the early diagnosis of IPA, the CT findings are not 100% sensitive nor specific [74,77]. Some CT signs have demonstrated excellent sensitivity and specificity in select populations. The presence of a macronodule and/or consolidation or mass in the study by Park et al. had a sensitivity of 98% for IPA, while the presence of the hypodensity sign had excellent specificity for IA in the study by Horger et al., approaching a 100% offset by a low specificity of 30% [78,79]. However, most of these studies focus on organ transplant recipients or otherwise immunocompromised patients. The "halo" sign is important among neutropenic patients, but nonspecific for the diagnosis of IPA in other groups of patients [21]. No CT findings are specific enough to reliably differentiate IA in ICU patients where a significant overlap between the radiological features of the co-existing conditions, such as ARDS, COPD, atelectasis, etc., is present [7,12,21]. The overall sensitivity and specificity of CT imaging is comparable to that of thoracic MRI; however, it is far more widely available [74].

PET/CT can add value to anatomy-based studies, for example, CT and CXR. It can assist in detecting lesions outside the limited area that are included in traditional anatomy-based imaging, which is important in case of disseminated infections. Furthermore, serial images can aid in monitoring the response to treatment and aid in the decision to continue or stop antifungals [80–85].

7. Treatment of Invasive Aspergillosis

Currently, the triazoles, voriconazoles and isavuconazole are considered first-line treatment options for invasive aspergillosis in patients with hematological malignancies as per the ESCMID-ECMM-ERS guideline [86]. The IDSA guidelines, published in 2016, name voriconazole as the primary therapy of choice for most of the invasive aspergillus infections, whereas isavuconazole and liposomal amphotericin B are considered alternative therapies (Table 5).

Table 5. Summary of recommendations for the treatment of Aspergillosis (from the IDSA-2016 guidelines [87]).

	Therapy		
Condition	**Primary**	**Alternative**	**Comments**
IPA	Voriconazole (6 mg/kg IV every 12 h for 1 d, followed by 4 mg/kg IV every 12 h; oral therapy can be used at 200–300 mg every 12 h or weight-based dosing on a mg/kg basis); see text for pediatric dosing	Primary: Liposomal AmB (3–5 mg/kg/day IV), isavuconazole 200 mg every 8 h for 6 doses, then 200 mg daily. Salvage: ABLC (5 mg/kg/day IV), caspofungin (70 mg/day IV × 1, then 50 mg/day IV thereafter), micafungin (100–150 mg/day IV), posaconazole (oral suspension: 200 mg TID; tablet: 300 mg BID on day 1, then 300 mg daily; IV: 300 mg BID on day 1, then 300 mg daily), itraconazole suspension (200 mg PO every 12 h)	Primary combination therapy is not routinely recommended; the addition of another agent or the switch to another drug class for salvage therapy may be considered in individual patients; dosage in pediatric patients for voriconazole and for caspofungin is different than that of adults; limited clinical experience is reported with anidulafungin; the dosage of posaconazole in pediatric patients has not been defined

Table 5. Cont.

Condition	Therapy: Primary	Therapy: Alternative	Comments
Invasive sinus aspergillosis	Similar to IPA	Similar to IPA	Surgical debridement as an adjunct to medical therapy
Tracheobronchial aspergillosis	Similar to IPA	Adjunctive inhaled AmB may be useful	Similar to IPA
Aspergillosis of the CNS	Similar to IPA	Similar to IPA Surgical resection may be beneficial in selected cases	This infection is associated with the highest mortality among all of the different patterns of IA; drug interactions with anticonvulsant therapy
Aspergillus infections of the heart (endocarditis, pericarditis and myocarditis)	Similar to IPA	Similar to IPA	Endocardial lesions caused by the Aspergillus species require surgical resection. Aspergillus pericarditis usually requires pericardiectomy
Aspergillus osteomyelitis and septic arthritis	Similar to IPA	Similar to IPA	Surgical resection of devitalized bone and cartilage is important for curative intent
Aspergillus peritonitis	Similar to IPA	Similar to IPA	Removal of peritoneal dialysis catheter is essential
Empiric and preemptive antifungal therapy	For empiric antifungal therapy, Liposomal AmB (3 mg/kg/day IV), caspofungin (70 mg day 1 IV and 50 mg/day IV thereafter), micafungin (100 mg day), voriconazole (6 mg/kg IV every 12 h for 1 day, followed by 4 mg/kg IV every 12 h; oral therapy can be used at 200–300 mg every 12 h or 3–4 mg/kg q 12 h)		Pre-emptive therapy is a logical extension of empiric antifungal therapy in defining a high-risk population with evidence of invasive fungal infection (e.g., pulmonary infiltrate or positive GM assay result)

Abbreviations: AML—acute myelogenous leukemia; ARF—acute respiratory failure; BALF—bronchoalveolar lavage fluid; CAPA—COVID-19-associated pulmonary aspergillosis; HM—hematological malignancy; HSCT—hematopoietic stem cell transplant; IFD—invasive fungal disease; mcfDNA—microbial cell-free DNA; mNGS—metagenomic next-generation sequencing; RUO-BAL KT—research-use-only Karius test on bronchoalveolar lavage fluid; SOT—solid organ transplant.

Isavuconazole was shown to be non-inferior to voriconazole in the study by Maertens et al. and was noted to have fewer adverse events [88]. Alternative approaches include liposomal amphotericin B and itraconazole. Although combinations of drugs from different classes have been utilized frequently, randomized studies have not been conducted [89]. The combination of voriconazole plus low-dose anidulafungin revealed in vitro synergy against azole-resistant *A. fumigatus* [90]. The superiority of combination therapy over monotherapy has not been demonstrated but can be considered as salvage therapy in patients not responding to monotherapy [89,91–93]. Among echinocandins, micafungin and caspofungin are listed as salvage treatments of IA; however, caspofungin performance was suboptimal [94,95].

Voriconazole is considered the first-line option for patients without hematological malignancies as its use is associated with lower mortality [96,97]. Resistance to antifungals is an increasing global problem [98]. *Aspergillus* species can be intrinsically resistant to antifungals and can also develop secondary resistance during treatment [99,100]. According to CDC, about 7% of resistant isolates have been identified in the US, predominantly from patients with hematological malignancies and SOT recipients. The resistance to azoles among isolates of *A. fumigatus* in Europe continues to increase and poses a significant management challenge [99,101]. Resistance to azoles translates into higher mortality [102].

Antifungal susceptibility testing is recommended on clinically relevant Aspergillus isolates, especially when there is a lack of response to treatment and in patient groups or areas with known high levels of azole resistance [86]. IDSA recommends against routine antifungal susceptibility testing, unless there is a lack of response to treatment or for epidemiological purposes [87]. Some species are intrinsically resistant to certain antifungals (i.e., *A. terreus* is intrinsically resistant to amphotericin B, etc.), so species identification is necessary to guide treatment decisions [100]. MIC testing to determine susceptibility to azoles is carried out for resistance surveillance and clinical management. Local resistance rates guide the antifungal agent selection; if the local azole resistance rate exceeds 10%, the ESCMID-ECMM-ERS guideline recommends adding echinocandin or using liposomal amphotericin B [86]. A similar approach is recommended for isolates with high MICs to voriconazole [86].

Therapeutic drug monitoring is recommended for itraconazole, voriconazole, posaconazole and flucytosine to ensure both the safety and efficacy of the therapy [86]. There is limited data to routinely support isavuconazole therapeutic drug monitoring; however, it can be considered when there are concerns regarding treatment failure or drug interactions [103,104]. A study by Mikulska et al. suggested that significantly lower serum levels of isavuconazole were observed among ICU vs. non-ICU patients, which may necessitate therapeutic drug monitoring of isavuconazole in the ICU patients to ensure efficacy [103].

Novel antifungals in the pipeline have differing mechanisms of action and are being actively studied as treatment options for invasive aspergillosis [105,106]. Fosmanogepix is a first-in-class antifungal drug that inhibits the fungal glycosylphosphatidylinositol (GPI)-anchored wall-transfer protein 1 (Gwt1 protein), which affects the maturation and localization of the fungal wall mannoproteins, impacting the cell wall integrity and disrupting the processes of hyphal formation and biofilm formation. GPI has shown promise in azole-resistant aspergillosis and has a favorable side-effect profile [107–111]. The combination of fosmanogepix and Liposomal amphotericin B has shown promising results in in vitro models [112].

Opelconazole is a long-acting inhaled triazole, which achieves high concentrations in the respiratory tract and has limited systemic absorption which reduces adverse events and drug–drug interactions [113]. Olorofim, a reversible inhibitor dihydroorotate dehydrogenase, is active against azole- and amphotericin B-resistant *Aspergillus* species, including some of the non-*fumigatus Aspergillus* species [95,114,115]. A phase 3 study comparing Olorofim versus liposomal amphotericin B, followed by standard of care therapy (OASIS) in the management of invasive aspergillosis of the lower respiratory tract disease, is in progress (NCT05101187).

Ibrexafungerp, another newer antifungal, works by inhibiting (1,3)-β-D-glucan synthase in *Aspergillus*, *Candida* and other fungi [116]. It has demonstrated potent activity in vitro against *Aspergillus* species and demonstrated synergistic activity with voriconazole, isavuconazole and amphotericin B [117,118]. Finally, rezafungin, is a long-acting echinocandin, approved for candidemia and invasive candidiasis. It has demonstrated in vitro activity against *Aspergillus* spp, including azole-resistant species. However, it has not been studied in invasive aspergillosis and further studies are necessary [119–121].

8. Prophylaxis

Antifungal chemoprophylaxis is a standard of care in patients with prolonged neutropenia, acute myeloid leukemia or MDS, history of invasive aspergillosis per engraftment, GVHD after a follow-up SCT, rare cases of inherited immunodeficiency and select SOT. Prophylactic posaconazole reduces the incidence of invasive aspergillosis among patients with prolonged neutropenia and stem cell transplant with COVID or AML. The value of prophylaxis to reduce the incidence of IA in non-neutropenic patients has not been established.

Currently, antifungals are not indicated for the prevention of IA in the ICU setting. Studies indicated that antifungal prophylaxis might be beneficial if the baseline incidence of IA is greater than 15–30%.

The Isavu-CAPA trial, investigating the use of isavuconazole for the prevention of CAPA was terminated early due to participant enrollment challenges [122]. According to the observational study by Hatzl et al., antifungal prophylaxis among ICU patients with COVID-19 was associated with significantly reduced CAPA incidence; however, it did not result in improved survival [123].

In 2021, Vanderbeke et al. published the results of a randomized, open-label, proof-of-concept trial evaluating posaconazole for the prevention of invasive pulmonary aspergillosis and critically ill influenza patients (POSA-FLU). Seven days of posaconazole prophylaxis were compared to no prophylaxis (standard of care, SOC) in these subjects. The incidence of IPA was not significantly reduced in the treatment arm when compared to the standard of care. However, this study was underpowered because most of the patients with IPA were excluded from the study as IPA was diagnosed within 48 h of ICU admission [124].

The retrospective study evaluating nebulized conventional amphotericin B as a means of prophylaxis of CAPA also showed a decreased incidence of CAPA in the prophylaxis groups; however, the mortality benefit was not demonstrated [125,126].

9. Conclusions

The diagnosis and management of IA in the ICU setting, despite it being more widely recognized by clinical providers, remains challenging. It contributes to an increase in mortality and poorer outcomes among patients in the ICU. It remains an under-recognized infection and awareness among providers remains low, contributing to the poor overall outcomes. Furthermore, variations in the definitions of IA in non-neutropenic patients further contribute to the insufficiently accurate estimation of the true incidence and prevalence of IA in the ICU.

It is essential to raise awareness among critical care physicians regarding the incidence of IA among ICU patients. In particular, the absence of "classical risk factors" may aid in the early diagnosis and appropriate treatment of these infections, which should lead to improved outcomes. Further studies are necessary in the ICU to aid in developing appropriate clinical guidelines and protocols to help navigate the approach to diagnosis and treatment of IA among these patients.

Author Contributions: A.Z. and J.A.V. both contributed to manuscript. All authors have read and agreed to the published version of the manuscript.

Funding: This research received no external funding.

Data Availability Statement: No new data were created or analyzed in this study.

Conflicts of Interest: A.Z.: None; J.A.V. Consultant for Melinta, F2G, Scynexis, Basilea, the author declares no conflict of interest.

References

1. Gaffney, S.; Kelly, D.M.; Rameli, P.M.; Kelleher, E.; Martin-Loeches, I. Invasive pulmonary aspergillosis in the intensive care unit: Current challenges and best practices. *APMIS* **2023**, *131*, 654–667. [CrossRef] [PubMed]

2. Jenks, J.D.; Nam, H.H.; Hoenigl, M. Invasive aspergillosis in critically ill patients: Review of definitions and diagnostic approaches. *Mycoses* **2021**, *64*, 1002–1014. [CrossRef] [PubMed] [PubMed Central]
3. Bartoletti, M.; Pascale, R.; Cricca, M.; Rinaldi, M.; Maccaro, A.; Bussini, L.; Fornaro, G.; Tonetti, T.; Pizzilli, G.; Francalanci, E.; et al. Epidemiology of Invasive Pulmonary Aspergillosis Among Intubated Patients With COVID-19: A Prospective Study. *Clin. Infect. Dis.* **2021**, *73*, e3606–e3614. [CrossRef] [PubMed] [PubMed Central]
4. White, P.L.; Dhillon, R.; Cordey, A.; Hughes, H.; Faggian, F.; Soni, S.; Pandey, M.; Whitaker, H.; May, A.; Morgan, M.; et al. A National Strategy to Diagnose Coronavirus Disease 2019-Associated Invasive Fungal Disease in the Intensive Care Unit. *Clin Infect Dis.* **2021**, *73*, e1634–e1644. [CrossRef] [PubMed] [PubMed Central]
5. Wang, Y.; Yao, Y.; Zhang, Q.; Chen, H.; He, Y.; Hu, K. Clinical courses and outcomes of COVID-19 associated pulmonary aspergillosis in 168 patients with the SARS-CoV-2 omicron variant. *BMC Infect. Dis.* **2024**, *24*, 117. [CrossRef] [PubMed] [PubMed Central]
6. Baddley, J.W.; Stephens, J.M.; Ji, X.; Gao, X.; Schlamm, H.T.; Tarallo, M. Aspergillosis in Intensive Care Unit (ICU) patients: Epidemiology and economic outcomes. *BMC Infect. Dis.* **2013**, *13*, 29. [CrossRef] [PubMed] [PubMed Central]
7. Meersseman, W.; Lagrou, K.; Maertens, J.; Van Wijngaerden, E. Invasive aspergillosis in the intensive care unit. *Clin. Infect. Dis.* **2007**, *45*, 205–216. [CrossRef] [PubMed]
8. Vandewoude, K.H.; Blot, S.I.; Depuydt, P.; Benoit, D.; Temmerman, W.; Colardyn, F.; Vogelaers, D. Clinical relevance of Aspergillus isolation from respiratory tract samples in critically ill patients. *Crit. Care* **2006**, *10*, R31. [CrossRef] [PubMed] [PubMed Central]
9. Meersseman, W.; Vandecasteele, S.J.; Wilmer, A.; Verbeken, E.; Peetermans, W.E.; Van Wijngaerden, E. Invasive aspergillosis in critically ill patients without malignancy. *Am. J. Respir. Crit. Care Med.* **2004**, *170*, 621–625. [CrossRef] [PubMed]
10. Blot, S.I.; Taccone, F.S.; Van den Abeele, A.M.; Bulpa, P.; Meersseman, W.; Brusselaers, N.; Dimopoulos, G.; Paiva, J.A.; Misset, B.; Rello, J.; et al. A clinical algorithm to diagnose invasive pulmonary aspergillosis in critically ill patients. *Am. J. Respir. Crit. Care Med.* **2012**, *186*, 56–64, Erratum in *Am. J. Respir. Crit. Care Med.* **2012**, *186*, 808. [CrossRef] [PubMed]
11. Taccone, F.S.; Van den Abeele, A.M.; Bulpa, P.; Misset, B.; Meersseman, W.; Cardoso, T.; Paiva, J.A.; Blasco-Navalpotro, M.; De Laere, E.; Dimopoulos, G.; et al. Epidemiology of invasive aspergillosis in critically ill patients: Clinical presentation, underlying conditions, and outcomes. *Crit. Care* **2015**, *19*, 7. [CrossRef] [PubMed] [PubMed Central]
12. van de Veerdonk, F.L.; Kolwijck, E.; Lestrade, P.P.; Hodiamont, C.J.; Rijnders, B.J.; van Paassen, J.; Haas, P.J.; Oliveira Dos Santos, C.; Kampinga, G.A.; Bergmans, D.C.; et al. Influenza-Associated Aspergillosis in Critically Ill Patients. *Am. J. Respir. Crit. Care Med.* **2017**, *196*, 524–527. [CrossRef] [PubMed]
13. Calderón-Parra, J.; Mills-Sanchez, P.; Moreno-Torres, V.; Tejado-Bravo, S.; Romero-Sánchez, I.; Balandin-Moreno, B.; Calvo-Salvador, M.; Portero-Azorín, F.; García-Masedo, S.; Muñez-Rubio, E.; et al. COVID-19-associated pulmonary aspergillosis (CAPA): Risk factors and development of a predictive score for critically ill COVID-19 patients. *Mycoses* **2022**, *65*, 541–550; Erratum in *Mycoses* **2022**, *65*, 1068. https://doi.org/10.1111/myc.13523. [CrossRef] [PubMed] [PubMed Central]
14. Toda, M.; Benedict, K.; Jackson, B.R. Invasive Aspergillosis After Influenza and Other Viral Respiratory Infections Among Intensive Care Unit Patients in a Commercially Insured Population in the United States, 2013-2018. *Open Forum Infect. Dis.* **2021**, *8*, ofab091. [CrossRef] [PubMed] [PubMed Central]
15. Schwartz, I.S.; Friedman, D.Z.P.; Zapernick, L.; Dingle, T.C.; Lee, N.; Sligl, W.; Zelyas, N.; Smith, S.W. High Rates of Influenza-Associated Invasive Pulmonary Aspergillosis May Not Be Universal: A Retrospective Cohort Study from Alberta, Canada. *Clin. Infect. Dis.* **2020**, *71*, 1760–1763. [CrossRef] [PubMed]
16. Waldeck, F.; Boroli, F.; Zingg, S.; Walti, L.N.; Wendel-Garcia, P.D.; Conen, A.; Pagani, J.L.; Boggian, K.; Schnorf, M.; Siegemund, M.; et al. Higher risk for influenza-associated pulmonary aspergillosis (IAPA) in asthmatic patients: A Swiss multicenter cohort study on IAPA in critically ill influenza patients. *Influenza Other Respir. Viruses* **2023**, *17*, e13059. [CrossRef] [PubMed] [PubMed Central]
17. Permpalung, N.; Chiang, T.P.Y.; Avery, R.K.; Ostrander, D.; Datta, K.; Segev, D.L.; Durand, C.M.; Zhang, S.X.; Massie, A.B.; Marr, K.A. Coronavirus Disease 2019–Associated Pulmonary Aspergillosis: A Noninvasive Screening Model for Additional Diagnostics. *Open Forum Infect. Dis.* **2023**, *10*, ofad155. [CrossRef]
18. Prattes, J.; Wauters, J.; Giacobbe, D.R.; Salmanton-García, J.; Maertens, J.; Bourgeois, M.; Reynders, M.; Rutsaert, L.; Van Regenmortel, N.; Lormans, P.; et al. Risk factors and outcome of pulmonary aspergillosis in critically ill coronavirus disease 2019 patients-a multinational observational study by the European Confederation of Medical Mycology. *Clin Microbiol Infect.* **2022**, *28*, 580–587. [CrossRef] [PubMed] [PubMed Central]
19. Nucci, M.; Anaissie, E. Fungal infections in hematopoietic stem cell transplantation and solid-organ transplantation--focus on aspergillosis. *Clin. Chest Med.* **2009**, *30*, 295–306. [CrossRef] [PubMed]
20. Townsend, L.; Martin-Loeches, I. Invasive *Aspergillosis* in the Intensive Care Unit. *Diagnostics* **2022**, *12*, 2712. [CrossRef] [PubMed] [PubMed Central]

21. Donnelly, J.P.; Chen, S.C.; Kauffman, C.A.; [illegible]; Lockhart, S.R.; Groll, A.H.; et al. Revision and Update of the Consensus Definitions of Invasive Fungal Disease From the European Organization for Research and Treatment of Cancer and the Mycoses Study Group Education and Research Consortium. *Clin. Infect. Dis.* **2020**, *71*, 1367–1376. [CrossRef] [PubMed] [PubMed Central]
22. Bassetti, M.; Azoulay, E.; Kullberg, B.J.; Ruhnke, M.; Shoham, S.; Vazquez, J.; Giacobbe, D.R.; Calandra, T. EORTC/MSGERC Definitions of Invasive Fungal Diseases: Summary of Activities of the Intensive Care Unit Working Group. *Clin. Infect. Dis.* **2021**, *72* (Suppl. S2), S121–S127. [CrossRef] [PubMed]
23. Hamam, J.; Navellou, J.C.; Bellanger, A.P.; Bretagne, S.; Winiszewski, H.; Scherer, E.; Piton, G.; Millon, L.; Collaborative RESSIF Group. New clinical algorithm including fungal biomarkers to better diagnose probable invasive pulmonary aspergillosis in ICU. *Ann. Intensive Care* **2021**, *11*, 41. [CrossRef] [PubMed] [PubMed Central]
24. Murray, C.K.; Loo, F.L.; Hospenthal, D.R.; Cancio, L.C.; Jones, J.A.; Kim, S.H.; Holcomb, J.B.; Wade, C.E.; Wolf, S.E. Incidence of systemic fungal infection and related mortality following severe burns. *Burns* **2008**, *34*, 1108–1112. [CrossRef] [PubMed]
25. Schauwvlieghe, A.F.A.D.; Rijnders, B.J.A.; Philips, N.; Verwijs, R.; Vanderbeke, L.; Van Tienen, C.; Lagrou, K.; Verweij, P.E.; Van de Veerdonk, F.L.; Gommers, D.; et al. Invasive aspergillosis in patients admitted to the intensive care unit with severe influenza: A retrospective cohort study. *Lancet Respir. Med.* **2018**, *6*, 782–792. [CrossRef] [PubMed]
26. Huang, L.; Zhang, N.; Huang, X.; Xiong, S.; Feng, Y.; Zhang, Y.; Li, M.; Zhan, Q. Invasive pulmonary aspergillosis in patients with influenza infection: A retrospective study and review of the literature. *Clin. Respir. J.* **2019**, *13*, 202–211. [CrossRef] [PubMed]
27. Kamai, Y.; Chiang, L.Y.; Lopes Bezerra, L.M.; Doedt, T.; Lossinsky, A.S.; Sheppard, D.C.; Filler, S.G. Interactions of Aspergillus fumigatus with vascular endothelial cells. *Med. Mycol.* **2006**, *44* (Suppl. S1), S115–S117. [CrossRef] [PubMed]
28. Feys, S.; Carvalho, A.; Clancy, C.J.; Gangneux, J.P.; Hoenigl, M.; Lagrou, K.; Rijnders, B.J.A.; Seldeslachts, L.; Vanderbeke, L.; van de Veerdonk, F.L.; et al. Influenza-associated and COVID-19-associated pulmonary aspergillosis in critically ill patients. *Lancet Respir. Med.* **2024**, *12*, 728–742. [CrossRef] [PubMed]
29. Koehler, P.; Bassetti, M.; Chakrabarti, A.; Chen, S.C.A.; Colombo, A.L.; Hoenigl, M.; Klimko, N.; Lass-Flörl, C.; Oladele, R.O.; Vinh, D.C.; et al. Defining and managing COVID-19-associated pulmonary aspergillosis: The 2020 ECMM/ISHAM consensus criteria for research and clinical guidance. *Lancet Infect Dis.* **2021**, *21*, e149–e162. [CrossRef] [PubMed] [PubMed Central]
30. Qin, C.; Zhou, L.; Hu, Z.; Zhang, S.; Yang, S.; Tao, Y.; Xie, C.; Ma, K.; Shang, K.; Wang, W.; et al. Dysregulation of Immune Response in Patients With Coronavirus 2019 (COVID-19) in Wuhan, China. *Clin Infect Dis.* **2020**, *71*, 762–768. [CrossRef] [PubMed] [PubMed Central]
31. Hurt, W.; Youngs, J.; Ball, J.; Edgeworth, J.; Hopkins, P.; Jenkins, D.R.; Leaver, S.; Mazzella, A.; Molloy, S.F.; Schelenz, S.; et al. COVID-19-associated pulmonary aspergillosis in mechanically ventilated patients: A prospective, multicentre UK study. *Thorax* **2023**, *79*, 75–82. [CrossRef] [PubMed] [PubMed Central]
32. Kula, B.E.; Clancy, C.J.; Hong Nguyen, M.; Schwartz, I.S. Invasive mould disease in fatal COVID-19: A systematic review of autopsies. *Lancet Microbe* **2021**, *2*, e405–e414. [CrossRef] [PubMed] [PubMed Central]
33. Moni, M.; Sathyapalan, D.T.; Edathadathil, F.; Razak, M.A.; Nair, S.G.; Nair, C.V.; Samban, S.S.; Prasanna, P.; Kulirankal, K.G.; Purushothaman, S.S.; et al. Predicting COVID 19-Associated Pulmonary Aspergillosis Risk in Low- and Middle-Income Countries: A Matched Case-Control Study. *Open Forum Infect. Dis.* **2024**, *11*, ofae406. [CrossRef] [PubMed] [PubMed Central]
34. Iacovelli, A.; Oliva, A.; Mirabelli, F.M.; Giannone, S.; Laguardia, M.; Morviducci, M.; Nicolardi, M.L.; Repaci, E.; Sanzari, M.T.; Leanza, C.; et al. Risk factors for COVID-19 associated pulmonary aspergillosis and outcomes in patients with acute respiratory failure in a respiratory sub-intensive care unit. *BMC Infect. Dis.* **2024**, *24*, 392. [CrossRef] [PubMed] [PubMed Central]
35. Gioia, F.; Walti, L.N.; Orchanian-Cheff, A.; Husain, S. Risk factors for COVID-19-associated pulmonary aspergillosis: A systematic review and meta-analysis. *Lancet Respir. Med.* **2024**, *12*, 207–216; Erratum in *Lancet Respir. Med.* **2024**, *12*, e42. https://doi.org/10.1016/S2213-2600(24)00121-8. [CrossRef] [PubMed]
36. Cornillet, A.; Camus, C.; Nimubona, S.; Gandemer, V.; Tattevin, P.; Belleguic, C.; Chevrier, S.; Meunier, C.; Lebert, C.; Aupée, M.; et al. Comparison of epidemiological, clinical, and biological features of invasive aspergillosis in neutropenic and nonneutropenic patients: A 6-year survey. *Clin. Infect. Dis.* **2006**, *43*, 577–584. [CrossRef] [PubMed]
37. Liu, L.; Gu, Y.; Wang, Y.; Shen, K.; Su, X. The Clinical Characteristics of Patients With Nonneutropenic Invasive Pulmonary Aspergillosis. *Front. Med.* **2021**, *8*, 631461. [CrossRef] [PubMed] [PubMed Central]
38. Guarner, J.; Brandt, M.E. Histopathologic diagnosis of fungal infections in the 21st century. *Clin. Microbiol. Rev.* **2011**, *24*, 247–280. [CrossRef] [PubMed] [PubMed Central]
39. Guinea, J.; Torres-Narbona, M.; Gijón, P.; Muñoz, P.; Pozo, F.; Peláez, T.; de Miguel, J.; Bouza, E. Pulmonary aspergillosis in patients with chronic obstructive pulmonary disease: Incidence, risk factors, and outcome. *Clin. Microbiol. Infect.* **2010**, *16*, 870–877. [CrossRef] [PubMed]

40. Kaziani, K.; Mitrakou, E.; Dimopoulos, G. Improving diagnostic accuracy for invasive pulmonary aspergillosis in the intensive care unit. *Ann. Transl. Med.* **2016**, *4*, 332. [CrossRef] [PubMed] [PubMed Central]
41. Chun, J.Y.; Jeong, S.J.; Kim, S.; Choi, S.; Lee, J.H.; Chung, H.S.; Park, S.; Lee, H.; Kim, H.Y.; Hwangbo, B.; et al. Performance of the galactomannan test for the diagnosis of invasive pulmonary aspergillosis using non-invasive proximal airway samples. *J. Infect.* **2024**, *88*, 106159. [CrossRef] [PubMed]
42. Kimura, S.I.; Odawara, J.; Aoki, T.; Yamakura, M.; Takeuchi, M.; Matsue, K. Detection of sputum Aspergillus galactomannan for diagnosis of invasive pulmonary aspergillosis in haematological patients. *Int. J. Hematol.* **2009**, *90*, 463–470. [CrossRef] [PubMed]
43. Bergeron, A.; Porcher, R.; Sulahian, A.; de Bazelaire, C.; Chagnon, K.; Raffoux, E.; Vekhoff, A.; Cornet, M.; Isnard, F.; Brethon, B.; et al. The strategy for the diagnosis of invasive pulmonary aspergillosis should depend on both the underlying condition and the leukocyte count of patients with hematologic malignancies. *Blood* **2012**, *119*, 1831–1837. [CrossRef] [PubMed]
44. Maertens, J.; Maertens, V.; Theunissen, K.; Meersseman, W.; Meersseman, P.; Meers, S.; Verbeken, E.; Verhoef, G.; Van Eldere, J.; Lagrou, K. Bronchoalveolar lavage fluid galactomannan for the diagnosis of invasive pulmonary aspergillosis in patients with hematologic diseases. *Clin. Infect. Dis.* **2009**, *49*, 1688–1693. [CrossRef] [PubMed]
45. Marr, K.A.; Balajee, S.A.; McLaughlin, L.; Tabouret, M.; Bentsen, C.; Walsh, T.J. Detection of galactomannan antigenemia by enzyme immunoassay for the diagnosis of invasive aspergillosis: Variables that affect performance. *J. Infect. Dis.* **2004**, *190*, 641–649. [CrossRef] [PubMed]
46. Boonsarngsuk, V.; Niyompattama, A.; Teosirimongkol, C.; Sriwanichrak, K. False-positive serum and bronchoalveolar lavage Aspergillus galactomannan assays caused by different antibiotics. *Scand. J. Infect. Dis.* **2010**, *42*, 461–468. [CrossRef] [PubMed]
47. Asano-Mori, Y.; Kanda, Y.; Oshima, K.; Kako, S.; Shinohara, A.; Nakasone, H.; Kaneko, M.; Sato, H.; Watanabe, T.; Hosoya, N.; et al. False-positive Aspergillus galactomannan antigenaemia after haematopoietic stem cell transplantation. *J. Antimicrob. Chemother.* **2008**, *61*, 411–416. [CrossRef] [PubMed]
48. Wheat, L.J.; Walsh, T.J. Diagnosis of invasive aspergillosis by galactomannan antigenemia detection using an enzyme immunoassay. *Eur. J. Clin. Microbiol. Infect. Dis.* **2008**, *27*, 245–251. [CrossRef] [PubMed]
49. Cruciani, M.; Mengoli, C.; Barnes, R.; Donnelly, J.P.; Loeffler, J.; Jones, B.L.; Klingspor, L.; Maertens, J.; Morton, C.O.; White, L.P. Polymerase chain reaction blood tests for the diagnosis of invasive aspergillosis in immunocompromised people. *Cochrane Database Syst. Rev.* **2019**, *9*, CD009551. [CrossRef] [PubMed] [PubMed Central]
50. Egger, M.; Jenks, J.D.; Hoenigl, M.; Prattes, J. Blood Aspergillus PCR: The Good, the Bad, and the Ugly. *J. Fungi* **2020**, *6*, 18. [CrossRef] [PubMed] [PubMed Central]
51. Boch, T.; Reinwald, M.; Spiess, B.; Liebregts, T.; Schellongowski, P.; Meybohm, P.; Rath, P.M.; Steinmann, J.; Trinkmann, F.; Britsch, S.; et al. Detection of invasive pulmonary aspergillosis in critically ill patients by combined use of conventional culture, galactomannan, 1-3-beta-D-glucan and Aspergillus specific nested polymerase chain reaction in a prospective pilot study. *J. Crit. Care.* **2018**, *47*, 198–203. [CrossRef] [PubMed]
52. Imbert, S.; Meyer, I.; Palous, M.; Brossas, J.Y.; Uzunov, M.; Touafek, F.; Gay, F.; Trosini-Desert, V.; Fekkar, A. *Aspergillus* PCR in Bronchoalveolar Lavage Fluid for the Diagnosis and Prognosis of Aspergillosis in Patients With Hematological and Non-hematological Conditions. *Front. Microbiol.* **2018**, *9*, 1877. [CrossRef] [PubMed] [PubMed Central]
53. Mikulska, M.; Furfaro, E.; De Carolis, E.; Drago, E.; Pulzato, I.; Borghesi, M.L.; Zappulo, E.; Raiola, A.M.; Grazia, C.D.; Del Bono, V.; et al. Use of Aspergillus fumigatus real-time PCR in bronchoalveolar lavage samples (BAL) for diagnosis of invasive aspergillosis, including azole-resistant cases, in high risk haematology patients: The need for a combined use with galactomannan. *Med. Mycol.* **2019**, *57*, 987–996. [CrossRef] [PubMed] [PubMed Central]
54. Mikulska, M.; Furfaro, E.; Dettori, S.; Giacobbe, D.R.; Magnasco, L.; Dentone, C.; Ball, L.; Russo, C.; Taramasso, L.; Vena, A.; et al. Aspergillus-PCR in bronchoalveolar lavage—Diagnostic accuracy for invasive pulmonary aspergillosis in critically ill patients. *Mycoses* **2022**, *65*, 411–418. [CrossRef] [PubMed]
55. Eigl, S.; Hoenigl, M.; Spiess, B.; Heldt, S.; Prattes, J.; Neumeister, P.; Wolfler, A.; Rabensteiner, J.; Prueller, F.; Krause, R.; et al. Galactomannan testing and Aspergillus PCR in same-day bronchoalveolar lavage and blood samples for diagnosis of invasive aspergillosis. *Med. Mycol.* **2017**, *55*, 528–534. [CrossRef] [PubMed]
56. Madut, D.B.; Chemaly, R.F.; Dadwal, S.S.; Hill, J.A.; Lee, Y.J.; Haidar, G.; Luk, A.; Drelick, A.; Chin-Hong, P.V.; Benamu, E.; et al. Clinical Utility of Plasma Microbial Cell-Free DNA Sequencing Among Immunocompromised Patients With Pneumonia. *Open Forum Infect. Dis.* **2024**, *11*, ofae425. [CrossRef] [PubMed] [PubMed Central]
57. Huygens, S.; Schauwvlieghe, A.; Wlazlo, N.; Moors, I.; Boelens, J.; Reynders, M.; Chong, G.L.; Klaassen, C.H.W.; Rijnders, B.J.A. Diagnostic Value of Microbial Cell-free DNA Sequencing for Suspected Invasive Fungal Infections: A Retrospective Multicenter Cohort Study. *Open Forum Infect. Dis.* **2024**, *11*, ofae252. [CrossRef] [PubMed] [PubMed Central]

58. Mah, J.; Nicholas, V.; Tayyar, R.; Moreno, A.; Murugesan, K.; Budvytiene, I.; Banaei, N. Superior Accuracy of Aspergillus Plasma Cell-Free DNA Polymerase Chain Reaction Over Serum Galactomannan for the Diagnosis of Invasive Aspergillosis. *Clin. Infect. Dis.* **2023**, *77*, 1282–1290. [CrossRef] [PubMed]
59. Cai, X.; Sun, C.; Zhong, H.; Cai, Y.; Cao, M.; Wang, L.; Sun, W.; Tao, Y.; Ma, G.; Huang, B.; et al. The value of metagenomic next-generation sequencing with different nucleic acid extracting methods of cell-free DNA or whole-cell DNA in the diagnosis of non-neutropenic pulmonary aspergillosis. *Front. Cell Infect. Microbiol.* **2024**, *14*, 1398190. [CrossRef] [PubMed] [PubMed Central]
60. Hoenigl, M.; Egger, M.; Price, J.; Krause, R.; Prattes, J.; White, P.L. Metagenomic Next-Generation Sequencing of Plasma for Diagnosis of COVID-19-Associated Pulmonary Aspergillosis. *J. Clin. Microbiol.* **2023**, *61*, e0185922. [CrossRef] [PubMed] [PubMed Central]
61. Liu, Z.; Chen, H.; Chen, D.; Wu, X.; Xu, H.; Chen, P.; Wang, R.; Chen, Y. Metagenomic next-generation sequencing for the diagnosis of invasive pulmonary aspergillosis in type 2 diabetes mellitus patients. *Sci. Rep.* **2024**, *14*, 16618. [CrossRef]
62. Bao, S.; Song, H.; Chen, Y.; Zhong, C.; Tang, H. Metagenomic next-generation sequencing for the diagnosis of pulmonary aspergillosis in non-neutropenic patients: A retrospective study. *Front. Cell Infect. Microbiol.* **2022**, *12*, 925982. [CrossRef] [PubMed] [PubMed Central]
63. Lee, K.H.; Won, D.; Kim, J.; Lee, J.A.; Kim, C.H.; Kim, J.H.; Jeong, S.J.; Ku, N.S.; Choi, J.Y.; Yeom, J.S.; et al. Utility of Plasma Microbial Cell-Free DNA Whole-Genome Sequencing for Diagnosis of Invasive Aspergillosis in Patients With Hematologic Malignancy or COVID-19. *J. Infect. Dis.* **2023**, *228*, 444–452. [CrossRef] [PubMed]
64. Jenks, J.D.; Mehta, S.R.; Taplitz, R.; Law, N.; Reed, S.L.; Hoenigl, M. Bronchoalveolar lavage Aspergillus Galactomannan lateral flow assay versus Aspergillus-specific lateral flow device test for diagnosis of invasive pulmonary Aspergillosis in patients with hematological malignancies. *J. Infect.* **2019**, *78*, 249–259. [CrossRef] [PubMed]
65. Miceli, M.H.; Maertens, J. Role of Non-Culture-Based Tests, with an Emphasis on Galactomannan Testing for the Diagnosis of Invasive Aspergillosis. *Semin. Respir. Crit. Care Med.* **2015**, *36*, 650–661. [CrossRef] [PubMed]
66. Hsiao, H.H.; Liu, Y.C.; Wang, H.C.; Du, J.S.; Tang, S.H.; Yeh, T.J.; Hsieh, C.Y.; Gau, Y.C.; Ke, Y.L.; Chuang, T.M.; et al. Comparison of a novel lateral-flow device to galactomannan assay at different time periods for detections of invasive aspergillosis. *J. Formos. Med. Assoc.* **2022**, *121*, 2123–2129. [CrossRef] [PubMed]
67. Heldt, S.; Hoenigl, M. Lateral Flow Assays for the Diagnosis of Invasive Aspergillosis: Current Status. *Curr. Fungal Infect. Rep.* **2017**, *11*, 45–51. [CrossRef] [PubMed] [PubMed Central]
68. Mercier, T.; Dunbar, A.; Veldhuizen, V.; Holtappels, M.; Schauwvlieghe, A.; Maertens, J.; Rijnders, B.; Wauters, J. Point of care aspergillus testing in intensive care patients. *Crit. Care* **2020**, *24*, 642. [CrossRef] [PubMed] [PubMed Central]
69. Egger, M.; Horvath, A.; Prüller, F.; Fickert, P.; Finkelman, M.; Kriegl, L.; Grønbaek, H.; Møller, H.J.; Prattes, J.; Krause, R.; et al. Fungal translocation measured by serum 1,3-ß-D-glucan correlates with severity and outcome of liver cirrhosis-A pilot study. *Liver Int.* **2023**, *43*, 1975–1983. [CrossRef] [PubMed] [PubMed Central]
70. Hoenigl, M.; Pérez-Santiago, J.; Nakazawa, M.; de Oliveira, M.F.; Zhang, Y.; Finkelman, M.A.; Letendre, S.; Smith, D.; Gianella, S. (1→3)-β-d-Glucan: A Biomarker for Microbial Translocation in Individuals with Acute or Early HIV Infection? *Front. Immunol.* **2016**, *7*, 404. [CrossRef] [PubMed] [PubMed Central]
71. Hoenigl, M. Fungal Translocation: A Driving Force Behind the Occurrence of Non-AIDS Events? *Clin. Infect. Dis.* **2020**, *70*, 242–244. [CrossRef] [PubMed] [PubMed Central]
72. Marty, F.M.; Koo, S. Role of (1-->3)-beta-D-glucan in the diagnosis of invasive aspergillosis. *Med. Mycol.* **2009**, *47* (Suppl. S1), S233–S240. [CrossRef] [PubMed]
73. Kuhlman, J.E.; Fishman, E.K.; Burch, P.A.; Karp, J.E.; Zerhouni, E.A.; Siegelman, S.S. CT of invasive pulmonary aspergillosis. *AJR Am. J. Roentgenol.* **1988**, *150*, 1015–1020. [CrossRef] [PubMed]
74. Alexander, B.D.; Lamoth, F.; Heussel, C.P.; Prokop, C.S.; Desai, S.R.; Morrissey, C.O.; Baddley, J.W. Guidance on Imaging for Invasive Pulmonary Aspergillosis and Mucormycosis: From the Imaging Working Group for the Revision and Update of the Consensus Definitions of Fungal Disease from the EORTC/MSGERC. *Clin. Infect. Dis.* **2021**, *72* (Suppl. S2), S79–S88. [CrossRef] [PubMed]
75. Greene, R.E.; Schlamm, H.T.; Oestmann, J.W.; Stark, P.; Durand, C.; Lortholary, O.; Wingard, J.R.; Herbrecht, R.; Ribaud, P.; Patterson, T.F.; et al. Imaging findings in acute invasive pulmonary aspergillosis: Clinical significance of the halo sign. *Clin. Infect. Dis.* **2007**, *44*, 373–379. [CrossRef] [PubMed]
76. Greene, R. The radiological spectrum of pulmonary aspergillosis. *Med. Mycol.* **2005**, *43* (Suppl. S1), S147–S154. [CrossRef] [PubMed]
77. Desoubeaux, G.; Bailly, É.; Chandenier, J. Diagnosis of invasive pulmonary aspergillosis: Updates and recommendations. *Med. Mal. Infect.* **2014**, *44*, 89–101. [CrossRef] [PubMed]

78. Park, S.Y.; Lim, C.; Lee, S.O.; Choi, S.H.; Kim, Y.S.; Woo, J.H.; Song, J.W.; Kim, M.Y.; Chae, E.J.; Do, K.H.; et al. Computed tomography findings in invasive pulmonary aspergillosis in non-neutropenic transplant recipients and neutropenic patients, and their prognostic value. *J. Infect.* **2011**, *63*, 447–456. [CrossRef] [PubMed]
79. Horger, M.; Einsele, H.; Schumacher, U.; Wehrmann, M.; Hebart, H.; Lengerke, C.; Vonthein, R.; Claussen, C.D.; Pfannenberg, C. Invasive pulmonary aspergillosis: Frequency and meaning of the "hypodense sign" on unenhanced, C. *T. Br. J. Radiol.* **2005**, *78*, 697–703. [CrossRef] [PubMed]
80. Ankrah, A.O.; Creemers-Schild, D.; de Keizer, B.; Klein, H.C.; Dierckx, R.A.J.O.; Kwee, T.C.; Span, L.F.R.; de Jong, P.A.; Sathekge, M.M.; Glaudemans, A.W.J.M. The Added Value of [^{18}F]FDG PET/CT in the Management of Invasive Fungal Infections. *Diagnostics* **2021**, *11*, 137. [CrossRef] [PubMed] [PubMed Central]
81. Sehgal, I.S.; Arora, K.; Agarwal, R.; Kumar, R.; Rana, N.; Dhooria, S.; Muthu, V.; Prasad, K.T.; Garg, M.; Rudramurthy, S.M.; et al. Role of serial fluorodeoxyglucose positron emission tomography-computed tomography (^{18}FDG-PET-CT) in assessing treatment response in treatment naïve chronic pulmonary aspergillosis subjects. *J. Infect. Dis.* **2024**, *19*, jiae409. [CrossRef] [PubMed]
82. Sharma, P.; Mukherjee, A.; Karunanithi, S.; Bal, C.; Kumar, R. Potential role of ^{18}F-FDG PET/CT in patients with fungal infections. *AJR Am. J. Roentgenol.* **2014**, *203*, 180–189. [CrossRef] [PubMed]
83. Hot, A.; Maunoury, C.; Poiree, S.; Lanternier, F.; Viard, J.P.; Loulergue, P.; Coignard, H.; Bougnoux, M.E.; Suarez, F.; Rubio, M.T.; et al. Diagnostic contribution of positron emission tomography with [^{18}F]fluorodeoxyglucose for invasive fungal infections. *Clin. Microbiol. Infect.* **2011**, *17*, 409–417. [CrossRef] [PubMed]
84. Ankrah, A.O.; Span, L.F.R.; Klein, H.C.; de Jong, P.A.; Dierckx, R.A.J.O.; Kwee, T.C.; Sathekge, M.M.; Glaudemans, A.W.J.M. Role of FDG PET/CT in monitoring treatment response in patients with invasive fungal infections. *Eur. J. Nucl. Med. Mol. Imaging* **2019**, *46*, 174–183. [CrossRef] [PubMed] [PubMed Central]
85. Altini, C.; Ruta, R.; Mammucci, P.; Rubini, D.; Ferrari, C.; Sardaro, A.; Pisani, A.R. Heterogeneous imaging features of Aspergillosis at ^{18}F-FDG PET/CT. *Clin. Transl. Imaging* **2022**, *10*, 435–445. [CrossRef]
86. Ullmann, A.J.; Aguado, J.M.; Arikan-Akdagli, S.; Denning, D.W.; Groll, A.H.; Lagrou, K.; Lass-Flörl, C.; Lewis, R.E.; Munoz, P.; Verweij, P.E.; et al. Diagnosis and management of Aspergillus diseases: Executive summary of the 2017 ESCMID-ECMM-ERS guideline. *Clin Microbiol Infect.* **2018**, *24* (Suppl. S1), e1–e38. [CrossRef] [PubMed]
87. Patterson, T.F.; Thompson, G.R., 3rd; Denning, D.W.; Fishman, J.A.; Hadley, S.; Herbrecht, R.; Kontoyiannis, D.P.; Marr, K.A.; Morrison, V.A.; Nguyen, M.H.; et al. Practice Guidelines for the Diagnosis and Management of Aspergillosis: 2016 Update by the Infectious Diseases Society of America. *Clin. Infect. Dis.* **2016**, *63*, e1–e60. [CrossRef] [PubMed] [PubMed Central]
88. Maertens, J.A.; Raad, I.I.; Marr, K.A.; Patterson, T.F.; Kontoyiannis, D.P.; Cornely, O.A.; Bow, E.J.; Rahav, G.; Neofytos, D.; Aoun, M.; et al. Isavuconazole versus voriconazole for primary treatment of invasive mould disease caused by Aspergillus and other filamentous fungi (SECURE): A phase 3, randomised-controlled, non-inferiority trial. *Lancet* **2016**, *387*, 760–769. [CrossRef] [PubMed]
89. Marr, K.A.; Schlamm, H.T.; Herbrecht, R.; Rottinghaus, S.T.; Bow, E.J.; Cornely, O.A.; Heinz, W.J.; Jagannatha, S.; Koh, L.P.; Kontoyiannis, D.P.; et al. Combination antifungal therapy for invasive aspergillosis: A randomized trial. *Ann. Intern. Med.* **2015**, *162*, 81–89; Erratum in *Ann. Intern. Med.* **2015**, *162*, 463. https://doi.org/10.7326/L15-5068; Erratum in *Ann. Intern. Med.* **2019**, *170*, 220. https://doi.org/10.7326/L18-0712. [CrossRef] [PubMed]
90. Siopi, M.; Siafakas, N.; Vourli, S.; Mouton, J.W.; Zerva, L.; Meletiadisc, J. Dose optimization of voriconazole/anidulafungin combination against Aspergillus fumigatus using an in vitro pharmacokinetic/pharmacodynamic model and response surface analysis: Clinical implications for azole-resistant aspergillosis. *J. Antimicrob. Chemother.* **2016**, *71*, 3135–3147. [CrossRef] [PubMed]
91. Panackal, A.A.; Parisini, E.; Proschan, M. Salvage combination antifungal therapy for acute invasive aspergillosis may improve outcomes: A systematic review and meta-analysis. *Int. J. Infect. Dis.* **2014**, *28*, 80–94. [CrossRef] [PubMed] [PubMed Central]
92. Raad, I.I.; Zakhem, A.E.; Helou, G.E.; Jiang, Y.; Kontoyiannis, D.P.; Hachem, R. Clinical experience of the use of voriconazole, caspofungin or the combination in primary and salvage therapy of invasive aspergillosis in haematological malignancies. *Int. J. Antimicrob. Agents* **2015**, *45*, 283–288. [CrossRef] [PubMed]
93. Spitzer, M.; Robbins, N.; Wright, G.D. Combinatorial strategies for combating invasive fungal infections. *Virulence* **2017**, *8*, 169–185. [CrossRef] [PubMed] [PubMed Central]
94. Maertens, J.; Raad, I.; Petrikkos, G.; Boogaerts, M.; Selleslag, D.; Petersen, F.B.; Sable, C.A.; Kartsonis, N.A.; Ngai, A.; Taylor, A.; et al. Caspofungin Salvage Aspergillosis Study Group. Efficacy and safety of caspofungin for treatment of invasive aspergillosis in patients refractory to or intolerant of conventional antifungal therapy. *Clin. Infect. Dis.* **2004**, *39*, 1563–1571. [CrossRef] [PubMed]
95. Egger, M.; Bellmann, R.; Krause, R.; Boyer, J.; Jakšić, D.; Hoenigl, M. Salvage Treatment for Invasive Aspergillosis and Mucormycosis: Challenges, Recommendations and Future Considerations. *Infect. Drug Resist.* **2023**, *16*, 2167–2178. [CrossRef] [PubMed] [PubMed Central]

96. Patel, D.A.; Gao, X.; Stephens, J.M.; Forshag, M.S.; Tarallo, M. US hospital database analysis of invasive aspergillosis in the chronic obstructive pulmonary disease non-traditional host. *J. Med. Econ.* **2011**, *14*, 227–237. [CrossRef] [PubMed]
97. Barchiesi, F.; Mazzocato, S.; Mazzanti, S.; Gesuita, R.; Skrami, E.; Fiorentini, A.; Singh, N. Invasive aspergillosis in liver transplant recipients: Epidemiology, clinical characteristics, treatment, and outcomes in 116 cases. *Liver Transpl.* **2015**, *21*, 204–212. [CrossRef] [PubMed]
98. Bosetti, D.; Neofytos, D. Invasive Aspergillosis and the Impact of Azole-resistance. *Curr. Fungal Infect. Rep.* **2023**, *17*, 77–86. [CrossRef] [PubMed] [PubMed Central]
99. De Francesco, M.A. Drug-Resistant *Aspergillus* spp.: A Literature Review of Its Resistance Mechanisms and Its Prevalence in Europe. *Pathogens.* **2023**, *12*, 1305. [CrossRef] [PubMed] [PubMed Central]
100. Van Der Linden, J.W.; Warris, A.; Verweij, P.E. Aspergillus species intrinsically resistant to antifungal agents. *Med. Mycol.* **2011**, *49* (Suppl. S1), S82–S89. [CrossRef] [PubMed]
101. Chowdhary, A.; Sharma, C.; Meis, J.F. Azole-Resistant Aspergillosis: Epidemiology, Molecular Mechanisms, and Treatment. *J. Infect. Dis.* **2017**, *216* (Suppl. S3), S436–S444. [CrossRef] [PubMed]
102. Lestrade, P.P.; Bentvelsen, R.G.; Schauwvlieghe, A.F.A.D.; Schalekamp, S.; van der Velden, W.J.F.M.; Kuiper, E.J.; van Paassen, J.; van der Hoven, B.; van der Lee, H.A.; Melchers, W.J.G.; et al. Voriconazole Resistance and Mortality in Invasive Aspergillosis: A Multicenter Retrospective Cohort Study. *Clin. Infect. Dis.* **2019**, *68*, 1463–1471. [CrossRef] [PubMed]
103. Mikulska, M.; Melchio, M.; Signori, A.; Ullah, N.; Miletich, F.; Sepulcri, C.; Limongelli, A.; Giacobbe, D.R.; Balletto, E.; Russo, C.; et al. Lower blood levels of isavuconazole in critically ill patients compared with other populations: Possible need for therapeutic drug monitoring. *J. Antimicrob. Chemother.* **2024**, *79*, 835–845. [CrossRef] [PubMed]
104. Risum, M.; Vestergaard, M.B.; Weinreich, U.M.; Helleberg, M.; Vissing, N.H.; Jørgensen, R. Therapeutic Drug Monitoring of Isavuconazole: Serum Concentration Variability and Success Rates for Reaching Target in Comparison with Voriconazole. *Antibiotics* **2021**, *10*, 487. [CrossRef] [PubMed] [PubMed Central]
105. Rauseo, A.M.; Coler-Reilly, A.; Larson, L.; Spec, A. Hope on the Horizon: Novel Fungal Treatments in Development. *Open Forum Infect. Dis.* **2020**, *7*, ofaa016. [CrossRef] [PubMed] [PubMed Central]
106. Hoenigl, M.; Sprute, R.; Egger, M.; Arastehfar, A.; Cornely, O.A.; Krause, R.; Lass-Flörl, C.; Prattes, J.; Spec, A.; Thompson, G.R., 3rd; et al. The Antifungal Pipeline: Fosmanogepix, Ibrexafungerp, Olorofim, Opelconazole, and Rezafungin. *Drugs* **2021**, *81*, 1703–1729. [CrossRef] [PubMed] [PubMed Central]
107. Pfaller, M.A.; Huband, M.D.; Flamm, R.K.; Bien, P.A.; Castanheira, M. Antimicrobial activity of manogepix, a first-in-class antifungal, and comparator agents tested against contemporary invasive fungal isolates from an international surveillance programme (2018-2019). *J. Glob. Antimicrob. Resist.* **2021**, *26*, 117–127. [CrossRef] [PubMed]
108. Hodges, M.R.; Ople, E.; Wedel, P.; Shaw, K.J.; Jakate, A.; Kramer, W.G.; Marle, S.V.; van Hoogdalem, E.J.; Tawadrous, M. Safety and Pharmacokinetics of Intravenous and Oral Fosmanogepix, a First-in-Class Antifungal Agent, in Healthy Volunteers. *Antimicrob. Agents Chemother.* **2023**, *67*, e0162322. [CrossRef] [PubMed] [PubMed Central]
109. Camargo, J.F.; Jabr, R.; Anderson, A.D.; Lekakis, L.; Diaz-Paez, M.; Briski, L.M.; Raja, M.; Morris, M.I.; Komanduri, K.V.; Pereira, D. Successful Treatment of Disseminated Disease Due to Highly Resistant Aspergillus calidoustus with a Novel Antifungal Therapy. *Antimicrob. Agents Chemother.* **2022**, *66*, e0220621. [CrossRef] [PubMed] [PubMed Central]
110. Kline, A.; Lionakis, M.S. Case Commentary: Long-Term Fosmanogepix Use in a Transplant Recipient with Disseminated Aspergillosis Caused by Azole-Resistant Aspergillus calidoustus. *Antimicrob. Agents Chemother.* **2022**, *66*, e0236821. [CrossRef] [PubMed] [PubMed Central]
111. Shaw, K.J.; Ibrahim, A.S. Fosmanogepix: A Review of the First-in-Class Broad Spectrum Agent for the Treatment of Invasive Fungal Infections. *J. Fungi* **2020**, *6*, 239. [CrossRef] [PubMed] [PubMed Central]
112. Gebremariam, T.; Gu, Y.; Alkhazraji, S.; Youssef, E.; Shaw, K.J.; Ibrahim, A.S. The Combination Treatment of Fosmanogepix and Liposomal Amphotericin B Is Superior to Monotherapy in Treating Experimental Invasive Mold Infections. *Antimicrob. Agents Chemother.* **2022**, *66*, e0038022. [CrossRef] [PubMed] [PubMed Central]
113. Cass, L.; Murray, A.; Davis, A.; Woodward, K.; Albayaty, M.; Ito, K.; Strong, P.; Ayrton, J.; Brindley, C.; Prosser, J.; et al. Safety and nonclinical and clinical pharmacokinetics of PC945, a novel inhaled triazole antifungal agent. *Pharmacol. Res. Perspect.* **2021**, *9*, e00690. [CrossRef] [PubMed] [PubMed Central]
114. Escribano, P.; Gómez, A.; Reigadas, E.; Muñoz, P.; Guinea, J.; ASPEIN Study Group. In vitro activity of olorofim against Aspergillus fumigatus sensu lato clinical isolates: Activity is retained against isolates showing resistance to azoles and/or amphotericin B. *Clin. Microbiol. Infect.* **2022**, *28*, e7–e1291. [CrossRef] [PubMed]
115. Feuss, A.; Bougnoux, M.E.; Dannaoui, E. The Role of Olorofim in the Treatment of Filamentous Fungal Infections: A Review of In Vitro and In Vivo Studies. *J. Fungi* **2024**, *10*, 345. [CrossRef] [PubMed] [PubMed Central]

116. Angulo, D.A.; Alexander, B.; Rautemaa-Richardson, R.; Alastruey-Izquierdo, A.; Hoenigl, M.; Ibrahim, A.S.; Ghannoum, M.A.; King, T.R.; Azie, N.E.; Walsh, T.J. Ibrexafungerp, a Novel Triterpenoid Antifungal in Development for the Treatment of Mold Infections. *J. Fungi* **2022**, *8*, 1121. [CrossRef] [PubMed] [PubMed Central]
117. Ghannoum, M.; Long, L.; Larkin, E.L.; Isham, N.; Sherif, R.; Borroto-Esoda, K.; Barat, S.; Angulo, D. Evaluation of the Antifungal Activity of the Novel Oral Glucan Synthase Inhibitor SCY-078, Singly and in Combination, for the Treatment of Invasive Aspergillosis. *Antimicrob. Agents Chemother.* **2018**, *62*, e00244-18. [CrossRef] [PubMed] [PubMed Central]
118. Petraitis, V.; Petraitiene, R.; Katragkou, A.; Maung, B.B.W.; Naing, E.; Kavaliauskas, P.; Barat, S.; Borroto-Esoda, K.; Azie, N.; Angulo, D.; et al. Combination Therapy with Ibrexafungerp (Formerly SCY-078), a First-in-Class Triterpenoid Inhibitor of (1→3)-β-d-Glucan Synthesis, and Isavuconazole for Treatment of Experimental Invasive Pulmonary Aspergillosis. *Antimicrob. Agents Chemother.* **2020**, *64*, e02429-19. [CrossRef] [PubMed] [PubMed Central]
119. Wiederhold, N.P.; Locke, J.B.; Daruwala, P.; Bartizal, K. Rezafungin (CD101) demonstrates potent in vitro activity against Aspergillus, including azole-resistant Aspergillus fumigatus isolates and cryptic species. *J. Antimicrob. Chemother.* **2018**, *73*, 3063–3067. [CrossRef] [PubMed]
120. Pfaller, M.A.; Messer, S.A.; Rhomberg, P.R.; Jones, R.N.; Castanheira, M. Activity of a long-acting echinocandin, CD101, determined using CLSI and EUCAST reference methods, against Candida and Aspergillus spp., including echinocandin- and azole-resistant isolates. *J. Antimicrob. Chemother.* **2016**, *71*, 2868–2873. [CrossRef] [PubMed] [PubMed Central]
121. Wiederhold, N.P.; Najvar, L.K.; Jaramillo, R.; Olivo, M.; Wickes, B.L.; Catano, G.; Patterson, T.F. Extended-Interval Dosing of Rezafungin against Azole-Resistant Aspergillus fumigatus. *Antimicrob. Agents Chemother.* **2019**, *63*, e01165-19. [CrossRef] [PubMed] [PubMed Central]
122. Jenks, J.D.; Hoenigl, M.; Thompson, G.R. Study protocol: A randomized, double-blind, placebo-controlled trial of isavuconazole prophylaxis for the prevention of COVID-19-associated pulmonary aspergillosis. *Contemp. Clin. Trials Commun.* **2024**, *39*, 101310. [CrossRef] [PubMed] [PubMed Central]
123. Hatzl, S.; Reisinger, A.C.; Posch, F.; Prattes, J.; Stradner, M.; Pilz, S.; Eller, P.; Schoerghuber, M.; Toller, W.; Gorkiewicz, G.; et al. Antifungal prophylaxis for prevention of COVID-19-associated pulmonary aspergillosis in critically ill patients: An observational study. *Crit. Care* **2021**, *25*, 335. [CrossRef] [PubMed] [PubMed Central]
124. Vanderbeke, L.; Janssen, N.A.F.; Bergmans, D.C.J.J.; Bourgeois, M.; Buil, J.B.; Debaveye, Y.; Depuydt, P.; Feys, S.; Hermans, G.; Hoiting, O.; et al. Posaconazole for prevention of invasive pulmonary aspergillosis in critically ill influenza patients (POSA-FLU): A randomised, open-label, proof-of-concept trial. *Intensive Care Med.* **2021**, *47*, 674–686. [CrossRef] [PubMed] [PubMed Central]
125. Melchers, M.; van Zanten, A.R.H.; Heusinkveld, M.; Leeuwis, J.W.; Schellaars, R.; Lammers, H.J.W.; Kreemer, F.J.; Haas, P.J.; Verweij, P.E.; van Bree, S.H.W. Nebulized Amphotericin B in Mechanically Ventilated COVID-19 Patients to Prevent Invasive Pulmonary Aspergillosis: A Retrospective Cohort Study. *Crit. Care Explor.* **2022**, *4*, e0696. [CrossRef] [PubMed] [PubMed Central]
126. Van Ackerbroeck, S.; Rutsaert, L.; Roelant, E.; Dillen, K.; Wauters, J.; Van Regenmortel, N. Inhaled liposomal amphotericin-B as a prophylactic treatment for COVID-19-associated pulmonary aspergillosis/aspergillus tracheobronchitis. *Crit. Care* **2021**, *25*, 298. [CrossRef] [PubMed] [PubMed Central]

Journal of
Fungi

MDPI

Review

Invasive Candidiasis in the Intensive Care Unit: Where Are We Now?

Jose A. Vazquez *, Lissette Whitaker and Ana Zubovskaia

Division of Infectious Disease, Medical College of Georgia, Augusta University, 1120 15th Street, Augusta, GA 30912, USA; lwhitaker@augusta.edu (L.W.); azubovskaia@augusta.edu (A.Z.)
* Correspondence: jvazquez@augusta.edu; Tel.: +1-706-721-1244

Abstract: Invasive fungal infections in the intensive care unit (ICU) are not uncommon and most cases are caused by *Candida* species, specifically *Candida albicans*. However, recently, there has been an increase in non-*albicans Candida* spp. (*C. glabrata*; *C. parapsilosis*) causing invasive fungal infections. This has led to an increasing awareness of this infection due to the increase in documented antifungal resistance in many of these *Candida* species. In addition, manifestations of invasive candidiasis are often non-specific, and the diagnosis remains extremely challenging. Unfortunately, delays in antifungal therapy continue to hamper the morbidity; length of stay; and the mortality of these infections. Although the echinocandins are the drugs of choice in these infections, antifungal resistance among the non-*albicans* species (*C. glabrata*; *C. krusei*; *C. auris*; *C. parapsilosis*) is being observed more frequently. This has led to an increase in morbidity and mortality, specifically in critically ill patients. Overall, the diagnosis and management of invasive candidiasis in the ICU remain challenging. It is imperative that the critical care physician keeps this infection at the forefront of their differential diagnosis in order to decrease the mortality rate of these individuals. In this review, we discuss the current epidemiologic trends, diagnosis, and management of invasive candidiasis in the intensive care unit setting.

Keywords: *Candida*; candidemia; invasive candidiasis; antifungals

1. Introduction

Fungal infections continue to be a dominant concern in immunocompromised patients. Recently, the Centers for Disease Control (CDC) and the World Health Organization (WHO) categorized fungal infections into different threat categories [1–3]. The published priority pathogens are ranked into three separate categories: critical, high, and medium priority. Of these categories, 6 of the 19 pathogens belong to the genus *Candida* (Table 1).

These encompass a range of infections caused by numerous *Candida* species. The growing problem reflects the enormous increase in the pool of patients at risk and the increased opportunity for *Candida* species to colonize and invade tissues which are normally resistant to invasion. *Candida* species are true opportunistic pathogens that are able to exploit recent advances in medical care and gain access to the circulatory system and invade deeper tissues. As expected, *Candida* species specifically affect high-risk patients who are either critically ill or immunocompromised.

Table 1. Distribution of *Candida* species [4–6].

Candida Species	Isolation Rate
C. albicans	50–60%
C. glabrata	20–40%
C. parapsilosis	10–20%
C. tropicalis	6–12%
C. krusei	1–3%
C. guillermondii	
C. lusitaniae	
C. dubliniensis	
C. auris	

2. Epidemiology

Invasive candidiasis (IC) remains the most common cause of invasive fungal infections worldwide. In fact, 6 of the top 19 fungal infections worldwide are caused by *Candida* species. There are over 160 different *Candida* species in nature; however, 10 of these produce over 95% of all candidal infections in humans [4]. These include *C. albicans*, *Nakaseomyces glabrata* (*Candida glabrata*), *C. parapsilosis*, *C. tropicalis*, *C. krusei*, *C. lusitaniae*, *C. guillermondii*, *C. dubliniensis*, and *C. auris* (Table 1). However, over the past decade, we have seen a shift from the more susceptible *Candida* species (*C. albicans*) to the less susceptible or resistant non-*albicans Candida* species, especially *C. glabrata* and, more recently, *C. auris* [4,7]. Of these non-*albicans Candida* species, *C. glabrata*, *C. krusei*, *C. lusitaniae*, and *C. auris* are of greatest importance because of their intrinsic resistance or the development of secondary resistance to current antifungals [4,7]. Of these, *C. glabrata* is probably the greatest threat, since it is the second most commonly isolated *Candida* species and has demonstrated increasing resistance to fluconazole (20–40%). Overall, it is intrinsically less susceptible to azoles and polyenes. A second potential threat is *C. auris*, which was first described in 2009 and has been reported in over 25 countries [8–10]. It has several unique characteristics including multidrug resistance to all antifungals, including almost 100% resistance to fluconazole, 20–30% resistance to polyenes, and occasional resistance to echinocandins, as well as the ability to colonize patients and persist in the healthcare environment [8–10] (Table 2).

Numerous epidemiologic studies have shown that all *Candida* species can be transmitted from person to person and from environment to person [4,8–10]. This has been recently described in patients that develop infections due to *C. auris*, whereby environmental colonization by this organism leads to hospital-acquired infections. Prior studies evaluating the molecular epidemiology of candidal infections have also shown environment-to-patient transmission of identical strains of *C. albicans* and *C. glabrata* [11,12].

Clinical and autopsy studies have confirmed the significant increase in the incidence of candidemia and disseminated candidiasis. This increase is multifactorial and reflects both increased recognition as well as a growing population of high-risk patients. Furthermore, the increase in infection also reflects the improved survival of patients with underlying immunosuppression and multiple comorbidities.

Candidemia and disseminated candidiasis mortality rates have not improved much over the past two decades and remain in the 30–40% range. Furthermore, in neutropenic patients, the mortality rate approaches 100% [4]. In addition, the diagnosis of IC continues to be associated with [illegible] in length of stay (70 days vs. 40 days) when compared to patients with no evidence of IC.

Table 2. Relative in vitro susceptibility patterns of the *Candida* species [13–17].

	Fluconazole	Voriconazole	Echinocandins	Polyenes	Ibrexafungerp	Fosmanogepix
C. albicans	S	S	S	S	S	S
C. glabrata	variable	variable	S	S	S	S
C. parapsilosis	S	S	S	S	S	S
C. tropicalis	S	S	S	S	S	S
C. krusei	R	S	S	S		
C. lusitaniae	S	S	S	variable	S	S
C. auris	R	R	S	variable	S	S

3. Pathogenesis and Host Defenses

In most fungal infections, host defects play a major role in the development of candidiasis. The skin constitutes a highly effective, impermeable barrier to *Candida* spp. Disruption of the skin from wounds, burns, and intravenous catheters permits invasion by opportunistic colonizing organisms. Effective phagocytosis is the critical defense mechanism that prevents candidal deep-tissue invasion, thereby limiting candidemia and preventing disseminated infection [18].

The initial step in the development of candidal infections is colonization of the skin and the gastrointestinal tract. Once the GI tract mucosa is disrupted by either chemotherapy (mucositis), trauma, or sepsis, organisms penetrate the injured tissue and gain access to the lymphatics and the bloodstream [4,19,20].

Numerous studies have demonstrated the most common risk factors contributing to candidemia and disseminated candidiasis. These include central venous catheters, prolonged ICU stays (>3 days), neutropenia, broad-spectrum antimicrobials, total parenteral nutrition, major surgery (GI tract), uncontrolled diabetes mellitus, *Clostridium difficile* infections, solid organ transplants, continuous renal replacement therapy (CRRT), extracorporeal membrane oxygenation (ECMO), the development of acute renal failure with hemodialysis, mechanical ventilation (>3 days), and severe acute pancreatitis [4–6,21–23] (Table 3). Large multicenter studies examining the incidence of IC in the ICU have been conducted over the past decade. The reported rates of candidemia vary significantly between 3.5 and 16.5 per 1000 admissions. However, due to inter-center variability, the fact that most studies focused on candidemia only, and the fact that some encompassed cases that likely represented colonization rather than IC, evaluation of IC incidence trends in the ICU over time remains a challenge [24–27].

A subset of postsurgical patients, particularly those with recurrent gastrointestinal perforation, anastomotic leaks, or acute necrotizing pancreatitis, are uniquely at high risk for developing candidemia and IC [28,29]. Several different risk factors have been derived from multiple studies. However, many of these risk factors lack specificity and broad application that would render most ICU patients eligible for empirical antifungal therapy [4,6,22].

A systemic review and meta-analysis by Thomas-Ruddel et al. evaluated 34 different studies evaluating the most important risk factors for candidiasis. The study evaluated 29 possible risk factors for IC in the ICU, including demographic factors, comorbid conditions, and medical interventions [27]. The authors found that the risk factors associated with the highest rate of IC were broad-spectrum antibiotics (OR, 5.6; 95% CI, 3.6–8.8), blood transfusions (OR, 4.9; 95% CI, 1.5–16.0), *Candida* colonization (OR, 4.7; 95% CI, 1.6–14.3), central venous catheters (OR, 4.7; 95% CI, 2.7–8.1), and total parenteral nutrition (OR, 4.6; 95% CI, 3.3–6.3). However, interaction among the various risk factors is probably high (Table 3) [4,6,22].

Within the hospital setting, the departments with the highest rates of candidemia include ICUs, surgical units, trauma units, and neonatal ICUs. Moreover, 25–50% of all nosocomial candidemia occurs in ICUs. More recently, non-*albicans Candida* species have also increased significantly in ICUs due to the increased use of fluconazole.

Table 3. Common risk factors for Candidemia and invasive Candidiasis [4,22,23].

Breach in barrier defense	Host-related: Burns Mucositis GI perforation Pancreatitis Anastomotic leak Neutropenia
	Medical Interventions Performed: Central venous catheter TPN Abdominal surgery GI/liver/biliary surgery Urological intervention Renal replacement therapy Mechanical ventilation Urinary catheter
Antibiotic use	Broad-spectrum antibiotics > 72 h
Candida colonization	Multifocal *Candida* colonization of the skin or mucus membranes of the gastrointestinal and urogenital tracts
Environmental acquisition	*C. auris* transmission
Comorbidities	Bacteremia
	Sepsis
	Shock
	Cancer (hematologic > solid organ)
	Transplant (BMT > solid organ) diabetes mellitus
	Liver failure
	Renal failure
	Pulmonary diseases
	Hematological diseases
	HIV
	Neutropenia
Drugs	Corticosteroids
	Chemotherapy
	Anti-rejection/immunosuppressive
	Biologics
Genetics	SNPs at loci: CD58, LCE4A-C1, orf68 TAGAP (14)

4. Manifestations

Candidal infections can present in a wide spectrum of clinical syndromes depending on the type of infection and the degree of immunosuppression. Unfortunately, there are no characteristic manifestations of IC. It has been suggested by Rex JH et al. and others that candidemia and systemic candidiasis may be divided into four overlapping groups or syndromes: catheter-related candidiasis, acute disseminated candidiasis, chronic disseminated candidiasis, and deep-tissue infection [30,31]. Although hematogenous dissemination occurs at some stage of the evolution of all four syndromes, only the first two are frequently associated with positive blood cultures. Thus, the use of candidemia as a marker of IC results in the underestimation of the true incidence of IC.

The clinical manifestation of candidemia and IC can vary significantly, including fever alone, leukocytosis alone, the absence of any organ-specific manifestations, or a wide spectrum of manifestations, including culture-negative septic shock (Table 4). Frequently, manifestations of candidemia may be superimposed on those of the underlying pathology and are indistinguishable from those associated with bacterial infections [4,6,20,22].

Table 4. Markers of invasive candidiasis [4–6].

• Fever unresponsive to broad-spectrum antimicrobials (frequently the only manifestation);
• Prolonged intravascular catheters;
• Leukocytosis;
• Prolonged stay in the ICU;
• Septic shock with multi-organ failure;
• Macronodular skin lesions (10–20%);
• Candidal endophthalmitis (5–8%).

Dissemination to multiple organ systems may occur in association with candidemia, especially to the kidney, liver, spleen, skin (macronodular skin lesions ~10%), eyes (endophthalmitis) (5–10%), brain, CSF, and heart valves [6,22].

5. Diagnosis

For the diagnosis of IC, findings from laboratory studies are non-specific and lack sensitivity [32,33]. It is most imperative to realize that the early and rapid detection of IC is important to initiate appropriate antifungal therapy and thus reduce the mortality rate associated with invasive candidiasis [34].

Blood cultures are generally the initial approach to diagnose candidemia and candidiasis. However, even if positive, it can occasionally take several days before *Candida* is identified, species identification is performed, and the in vitro susceptibility data are gathered. Unfortunately, blood cultures are positive only in 40–70% of cases of candidiasis [32,33]. In fact, in a large retrospective autopsy published by Roosen et al. in 2000, the authors evaluated 100 adult patients in the MICU. Of these, 81% of the patients had a diagnosis made antemortem. However, of the diagnoses that were missed antemortem (36%), the only infectious process missed was systemic candidiasis (36%) [35].

In other instances of deep-seated candidiasis, tissue cultures are needed. The incidence of deep-seated candidiasis without concomitant candidemia in the ICU is unclear due to challenges in obtaining sterile specimens for microbiological confirmation. Intra-abdominal candidiasis (IAC) accounts for most deep-seated cases of IC, with ~ 30% occurring in the critical care unit [29,36]. Perforation, anastomotic leaks, repeat laparotomies, necrotizing pancreatitis and abdominal organ transplants increase risk; therefore, incidence is higher in surgical ICUs [36].

6. Blood and Tissue Cultures

The gold standard for diagnosing invasive candidiasis is blood cultures, even though their sensitivity is low and the time required for species identification usually exceeds 48 h. Tissue cultures require a surgical approach and are sometimes hard to perform in an already critically ill patient. However, more recently, interventional radiology with percutaneous abscess drainage has produced dramatic results. To overcome these issues, culture-based tests can be combined with non-culture-based assays [33,37,38] (Table 5). For example, using MALDI-TOF technology, once the organism is isolated, the time to identification can be shortened to less than 2 h.

Table 5. Non-culture- and culture-based diagnostic tests for invasive candidiasis [4,22,39–41].

Test	Turnaround Time	Result	Sensitivity	Specificity	Miscellaneous
Culture	1–4 days	Positive	40–70%	N/A	Permits the identification of species and in vitro susceptibility
B-D-glucan	1–2 h	>80 ng/L	90%	80%	Non-specific for candidiasis
T2MR Candida	1 h	Positive	91%	99%	Allows the direct detection of five *Candida* spp in whole blood; limit of detection: ~1.3 CFU/mL
Molecular methods • PCR; • mcfDNA.	1–2 h	Positive	95%	92%	None are FDA-approved for clinical practice

Approximately 50% of blood cultures remain negative despite a diagnosis of IC and, occasionally, a missed diagnosis of candidemia is also encountered for numerous reasons [32,33]. However, candidemia due to line-related infections or endovascular infections with no deep-seated candidiasis are most likely to result in a positive blood culture. Patients with candidemia plus deep-seated candidiasis may present with occasional positive blood cultures. On the other hand, patients with deep-seated candidiasis, most commonly in the setting of intra-abdominal perforation, rarely have positive blood cultures [32,33].

Recognizing the varied spectrum of *Candida* infections (candidemia vs. candidiasis or both) that clinicians might encounter will help with early treatment and performing an accurate diagnostic approach [4,22].

7. Non-Culture-Based Tests (NCBTs)

The potential of NCBTs is increasingly being utilized to assist in clinical decision making in order to guide the following: (1) the *initiation* of pre-emptive antifungal therapy (AFT); (2) the discontinuation or withholding of empirical AFT; (3) the monitoring of clinical improvements in patients with IC. Although they are somewhat sensitive, these tests are frequently non-specific and broad-spectrum. Unfortunately, most institutions do not carry these assays in-house and, thus, it may take several days for the results to be returned. This makes them difficult to use in real time to assess for active infection [42].

1,3-β-d-glucan (BDG) test is one of the most widely used NCBTs. The assay has shown promise in the diagnosis and treatment stratification of systemic candidiasis infections in the ICU [22,37,39]. It is important to note that a positive BDG test is suggestive of infection and should be used in combination with clinical manifestations in high-risk groups (Table 5).

In addition to monitoring qualitative BDG test results during interval testing, tracking quantitative values following the initiation of antifungal therapy may be used as a prognostic marker for patient response. However, the BDG test is a pan-fungal assay (it recognizes other fungi besides *Candida* species) with a sensitivity of 85% and a specificity of 40–90% [22,37,39].

A study conducted on 257 patients with proven IC that were treated with anidulafungin concluded that a decrease in BDG levels during therapy was associated with treatment success [40]. Consistently decreasing BDG levels during treatment have been shown by multiple groups to result in a favorable therapeutic response among patients with proven or probable invasive fungal infections.

Clinicians should also be wary of scenarios that may lead to false-positive BDG results. For example, certain medications can falsely elevate beta-D-glucan levels, including intravenous amoxicillin–clavulanate and piperacillin–tazobactam [38]. On occasion, infections by certain bacterial organisms can also lead to false positives including *S. pneumoniae* and *P. aeruginosa*, both of which also produce some degree of BDG. Another clinical scenario to be aware of is in hemodialysis patients since the cellulose filters used in dialysis may release BDG substrates that can lead to false positives [43,44]. Blood levels of BDG can also be elevated in the setting of fungal translocation via the GI tract due to a compromised intestinal epithelial barrier during episodes of mucositis or diarrhea due to immune dysfunction, gut damage, or altered gut microbiota [43,44].

The T2 magnetic resonance (T2MR) *Candida* assay (T2 Biosystems, Lexington, MA 02421, USA). is a rapid diagnostic test that uses magnetic resonance technology to directly detect and identify the presence of the five most common *Candida* species: *C. albicans*, *C. glabrate*, *C. parapsilosis*, *C. tropicalis*, and C. *krusei* [40]. The assay amplifies *Candida* DNA and detects the amplified fragments in whole blood using magnetic resonance properties, allowing for the faster detection of *Candida* in the blood compared to traditional culture methods [41,45,46] [Table 5].

In a study of 1801 subjects, T2MR Candida analyzed whole-blood samples and showed a sensitivity of 91.1% and a specificity of 99.4% with a turnaround time of 3 h [41].

There are several potential benefits when using the T2MR technology including the early diagnosis of candidemia, which can lead to improved therapeutic algorithms in invasive candidiasis. Furthermore, it can lead to increased case detection, species-specific tailored antifungal therapy, and a decrease in empiric antifungal treatment. This assay can play an adjunctive role as a stewardship tool to limit unnecessary antifungal utilization. It has been shown to have a high specificity and NPV when compared to blood cultures.

The BDG assay and T2MR Candida are now commonly used assays in the USA, where they have shown to be very helpful when deciding to discontinue empiric antifungal therapy due to their high NPV [47].

8. Multiplex Candida Real-Time PCR

The use of PCR in the diagnosis of fungal infections continues to be challenging. In fact, at the present time, for technical reasons, there are no commercially available tests. In a review of more than 50 standard, nested, or real-time PCRs, the pooled sensitivity and specificity were 95 and 92%, respectively, for invasive candidiasis [48]. Unfortunately, no PCR assays have been approved for commercial use in the US.

This test is expensive to perform, however, and although costs may be balanced by more accurate earlier diagnosis with reduced administration of antifungal agents and reduced mortality rates, further studies are needed to validate its use and assess its cost–benefit ratio [49].

9. Microbial Cell-Free DNA (mcfDNA) Sequencing

Microbial cell-free DNA (mcfDNA) represents a new frontier in infectious disease diagnostics that relies on the detection of fragments of pathogen DNA shed into the bloodstream from a distant site of infection. Detection can be achieved noninvasively through a blood draw; thus, the method has been referred to as a "liquid biopsy" [50]. This technique is often used as a final attempt in diagnosing infection, especially in immunocompromised and pediatric populations. Unlike traditional microbiological diagnostics, mcfDNA sequencing does not rely on organism growth and to some extent can detect a signal despite the presence of anti-infective treatments. In addition, it also allows for species identification across multiple kingdoms, including bacterial, viral (DNA viruses only), fungal, or parasitic organisms, without requiring prior knowledge of the specific organism [50]. This cell-free DNA (cfDNA) detection in plasma is a novel testing modality for the noninvasive diagnosis of invasive fungal infections (IFIs).

There are several limitations to these sensitive assays, including the leakage of molecular material from the microbiome that can be absorbed into the bloodstream and lead to a false positive. This is especially true in patients with acute inflammation, such as pneumonia. The detection of bacterial DNA in plasma might be particularly problematic in patients who are immunocompromised because it is very difficult to exclude the possibility of a true positive [51]. Furthermore, next-generation sequencing does not distinguish commensal or colonizing organisms from pathogenic organisms, so the results must be interpreted based on the clinical picture [52,53].

10. Management

The treatment of candidal infections varies considerably and is based on the anatomic location of infection, the patient's underlying disease and immune status, the species of *Candida*, and in some cases, the in vitro susceptibility of the different *Candida* species. In 2016, the Infectious Disease Society of America and the Mycosis Study Group published guidelines for the treatment of candidiasis [54] (Table 6). Several factors are involved in treating candidemia and IC; these include removing the focus of infection (catheter, abscess), removing or decreasing immunosuppression, restoring immune function, initiating early and appropriate antifungal therapy, and assessing for metastatic foci of infection (endophthalmitis) [4,22].

Although controversial, the need for an ophthalmologic exam is still recommended according to the Infectious Disease Society of America/Mycosis Study Group candidiasis guidelines [54]. In a recently published study by Lehman et al., the investigators found an ocular candidiasis incidence of 6% of patients [55]. In addition, they also found the highest sensitivity of a positive finding when the exams were performed at >7 days from the positive blood culture.

Table 6. Treatment of Candidemia and Invasive Candidiasis [54].

Antifungal Therapy	*Candida* Species			
	C. albicans *C. parapsilosis* *C. tropicalis*	*C. glabrata*	*C. krusei*	*C. auris*
Preferred Initial Therapy	Echinocandins	Echinocandins	Echinocandins	Echinocandins
Alternative Initial Therapy	Fluconazole	Liposomal AmB	Liposomal AmB or Voriconazole	Liposomal AmB
Step-Down Therapy	Fluconazole	Voriconazole	Voriconazole	Variable depending on in vitro susceptibility

The management of candidemia and IC in the ICU poses certain challenges. The presentation of IC in the ICU is frequently non-specific, the results of diagnostic tests are not readily available, and critically ill patients at the ICU are at risk of further rapid deterioration. These factors result in a common practice of empirically starting antifungal treatment. There are several studies that have shown that delays in the initiation of antifungal therapy translates to increased mortality among ICU patients. However, the results of randomized clinical trials have failed to demonstrate a clear benefit of pre-emptive empiric antifungal therapy [34,56,57]. The EMPIRICUS trial failed to demonstrate improved fungal infection-free survival among 251 non-neutropenic, critically ill patients with multiple *Candida* colonization, multiorgan failure, and antibiotic exposure, who were randomized to either empiric micafungin or placebo. On day 28, IFI-free survival was not statistically different between the micafungin and placebo groups [58]. However, there was a trend towards decreased rates of new proven IC in the micafungin group. In a separate study by Ostrosky-Zeichner et al., the multicenter, double-blinded, placebo-controlled MSG-01 trial randomized the patients at risk for IC in the ICU to either prophylactic caspofungin or placebo. No statistically significant difference was observed in terms of all-cause mortality as well as the development of proven IC [59]. A study by Knitsch et. al., INTENSE, aimed to assess the efficacy of pre-emptive antifungal therapy with micafungin vs. placebo among critically ill patients requiring surgery for intra-abdominal infections [60]. Again, no significant difference was seen in the occurrence of confirmed IC between the two arms, nor the median time to IC.

The most current IDSA guidelines provide a weak recommendation that fluconazole could be used among high-risk ICU patients for prophylaxis if the rate of invasive candidiasis in the ICU exceeds 5% [54]. The utility of empirical fluconazole for febrile patients in the ICU despite antibiotics was evaluated in a randomized, placebo-controlled trial by Schuster et al. [61]. Fluconazole failed to improve the composite outcome more than a placebo; however, admittedly, this study had several limitations.

It is also important to note that the widespread use of pre-emptive or prophylactic azole antifungal therapy in the ICU may be contributing to the selection of more resistant *Candida* species. According to a retrospective observational study from France, a significant increase in ICU-acquired *Candida glabrata* colonization was observed over the 2005–2012 study period, without a concomitant increase in mortality [62]. Another study detected a statistically significant correlation between the increase in caspofungin MICs among *C. parapsilosis* and *C. glabrata* and the increased use of caspofungin between 2007 and 2009 [63]. Other studies have shown that fluconazole-resistant *Candida* strains, for instance

C. parapsilosis, can thrive even in the absence of azole exposure, making it unclear if the restriction of azoles is enough to prevent the expansion of azole-resistant strains [64,65]. Nonetheless, antifungal stewardship programs are an essential part in the management of invasive candidiasis in the ICU to ensure the responsible use of antifungals. This includes inappropriate antifungal therapy in the form of overusing antifungals without an appropriate indication, incorrect dosing, incorrect choice of antifungal, or drug–drug interactions; these have been reported in multiple studies and might contribute to adverse outcomes among the ICU population [66–68].

The limitations of culture and non-culture-based methods used to make a diagnosis of IC contribute to excessive pre-emptive utilization of antifungals. Blood cultures are currently considered a gold standard; however, they are frequently negative in intra-abdominal candidiasis. Moreover, even in candidemia, the prolonged time to positivity negatively impacts the time to initiate appropriate therapy. The initiation of antifungal therapy based on the results of non-culture-based testing represents another potential strategy of antifungal therapy in the ICU [36,69]. Rapid diagnostic tests (MALDI-TOF, multiplex PCR, peptide nucleic acid fluorescent in situ hybridization, and T2 magnetic resonance detection panels) can improve the time to appropriate antifungal therapy and decrease the inappropriate utilization of antifungals [70–72]. Among these tests, the T2MR detection panel is the only test currently approved by the FDA that does not require prior incubation and growth on culture media and can be performed straight on a blood sample.

The antifungals that are currently utilized in the ICU setting for the treatment of invasive candidiasis belong to three major groups: echinocandins, azoles, and polyenes. According to the IDSA/MSG guidelines, echinocandins are the preferred agent of choice to initially treat unstable critically ill patients [13,54,73] (Table 6). In addition, echinocandins are also the preferred initial option for both neutropenic and non-neutropenic patients due to their ability to penetrate candidal biofilms, broader antifungal activity, favorable safety profile, good tolerance, and fewer significant drug–drug interactions as compared to azoles [54,74]. In a meta-analysis by Andes et al., initial antifungal therapy with echinocandins resulted in significantly improved 30-day and 90-day mortality rates as compared to fluconazole [14,75]. Fluconazole was shown to be a good option for de-escalation therapy. The IDSA guidelines suggest de-escalation after 5–7 days, after ensuring clinical stability. Current ESCMID guidelines recommend switching to oral therapy after 10 days of intravenous therapy [13,54]. Fluconazole can still be considered as an initial management option for candidemia among ICU patients without septic shock/multiorgan failure and in *Candida* species with no fluconazole resistance [14,54]. Notably, the rapidly emerging pathogen *C. auris* is almost universally resistant to fluconazole, making echinocandins the first-line treatment. However, resistance to all three major groups of antifungals has been documented among *C. auris* strains [76–78]. If fluconazole is selected as the initial antifungal management option, a loading dose should be administered. It is important to note that testing for antifungal susceptibility should be pursued in all clinically relevant *Candida* isolates [4,22,54]. Lipid formulations of amphotericin B are suggested as an alternative in cases when echinocandins and/or azoles are not available and in infections due to multidrug-resistant *Candida* strains. It is worth mentioning that *Candida* species have an ability to form biofilms and thus produce biofilm-associated infections, such as catheter-associated infections, which necessitate removal of the central venous catheter. If, however, for various reasons the catheter cannot be removed, echinocandins or liposomal AmB should be used as the first line. Although no RCTs are available, based on smaller studies and expert opinions, these antifungals can be of benefit due to their ability to penetrate candidal biofilms [74,79]. Most guidelines do not consider amphotericin B an attractive first-line option due to its inherent nephrotoxic potential. Voriconazole is recommended as a step-down option for *C. krusei* infections [54].

De-escalation of the antifungal therapy, preferably oral, seems an appropriate mode of action in clinically stable ICU patients after the recommended duration of parenteral referred therapy and in the absence of antifungal resistance. In the open-label trial by Vazquez et al., there was no difference when anidulafungin was de-escalated to fluconazole/voriconazole after five days of treatment vs. continued parenteral therapy throughout the duration of treatment if the prespecified criteria for de-escalation were met [80]. Post hoc analysis of the AmarCAND2 study also showed no association between systemic antifungal therapy de-escalation and an increase in 28-day mortality. However, the study population included patients with not only proven but also suspected candidiasis, with almost half of the de-escalation group having proven invasive candidiasis [81]. In the post hoc analysis by Moreno-Garcia et al., early de-escalation in candidemia caused by fluconazole-susceptible strains after source control had been achieved had no negative impact on 30-day mortality in the multivariable analysis of risk factors [82]. Notably, these two studies were not designed to compare the ongoing therapy with echinocandins and the de-escalation to azoles. Despite the published data regarding the safety of de-escalation in clinically stable patients with fluconazole-susceptible strains, the actual de-escalation rates vary. In fact, studies evaluating de-escalation reported levels of de-escalation of systemic antifungal therapy in the ICU as low as 20–23% [81–83]. Further antifungal stewardship efforts are necessary to ensure appropriate and timely de-escalation of systemic therapy [15].

Biomarker-driven de-escalation of systemic antifungal therapy has shown to be a reliable option among the ICU population. The BDG assay alone or in combination with other tests (Mn-Ag/Mn-Ab) has a good negative predictive value, providing evidence to rule out IC and discontinue the antifungal therapy, as shown in two RCTs [16,84].

As stated previously, source control remains a key element of the successful treatment of invasive candidiasis/candidemia in the ICU. Failure to achieve source control is associated with higher 30-day and hospital mortality [54,57,85].

The rate of invasive candidiasis caused by drug-resistant *Candida* strains is increasing concern, rendering some of the actively utilized agents, such as azoles and echinocandins, relatively ineffective. Several novel antifungals are currently being studied as potential medicines to overcome current challenges in the management of invasive candidiasis in the setting of antifungal resistance. Ibrexafungerp is an oral triterpenoid and works by inhibiting (1,3)-β-D-glucan synthase, similarly to echinocandins. It demonstrates potent in vitro activity against drug-resistant Candida species, including azole-resistant *Candida* spp., echinocandin-resistant *C. glabrata,* and MDR *C. auris* [86,87]. Currently, the utility of ibrexafungerp in invasive candidiasis and as step-down therapy in candidemia, including infections caused by MDR *C. auris,* is being evaluated in CARES and FURI studies, with promising interim results [17,87].

Fosmanogepix is another novel antifungal that has demonstrated potent activity against *C. auris* [87–89] in a phase-2 clinical trial evaluating fosmanogepix in invasive candidiasis and candidemia due to *C. auris* [88]. In addition, it also showed promise as a first-line treatment of candidemia in non-neutropenic patients, with a favorable side-effect profile [87–89].

Finally, rezafungin is a once-weekly echinocandin with a broad spectrum of activity against *Candida* spp., including *C. auris* [90,91]. Rezafungin has demonstrated excellent activity in treating both IC and candidemia in the two randomized clinical trials STRIVE and RESTORE [92,93]. However, its role in the initial management of ICU patients needs to be further established [92,93]. Rezafungin is currently being used to continue echinocandin therapy in cases when the patients are clinically stable and ready for discharge. This avoids the use of central venous catheters in the OPAT setting while continuing once-weekly echinocandin treatment.

11. Empiric Therapy

The empiric use of antifungal agents in febrile septic patients in the ICU is widespread despite the absence of significant clinical data. Given the existing data regarding the use of early and appropriate antifungal therapy and its correlation with decreased mortality, it appears reasonable to recommend empiric therapy in selected patients. It is important to note that the use of empiric antifungal therapy in any febrile, low-risk patient is not justified.

There are several studies that may be used for the early identification of patients that are at high risk of developing IC and thus may benefit from an antifungal, specifically, the *Candida* colonization index, the Candida score, and the Ostrosky-Zeichner clinical prediction rule [94–96].

The *Candida* colonization index (CCI) is based on the premise that most patients with IC have been previously colonized by *Candida* species. The development of invasive candidal infections is often preceded by extensive multifocal colonization of the skin or of the mucus membranes of the gastrointestinal and urogenital tracts, and it is dependent on the degree of colonization. This is assessed using the colonization index, which has been shown to be an independent risk factor for the development of candidemia and IC [94]. A 6-month prospective cohort study was carried out in patients admitted to the surgical and neonatal intensive care units in a 1600-bed university medical center. The authors concluded that the intensity of *Candida* colonization, which was assessed by systematic screening, helped in predicting subsequent infections with identical *Candida* strains in critically ill patients. This study identified high-risk patients with *Candida* colonization and may offer an opportunity for early intervention strategies [95]. However, its true applicability in critically ill patients is still limited without the use of more accurate predictors, such as specific biomarkers. The *Candida* colonization index remains an important way to characterize the dynamics of colonization, which increases early in patients who develop invasive candidiasis.

The *Candida* score is another method that may be used to identify high-risk patients prior to those patients developing positive blood cultures. In a large cohort of non-neutropenic, critically ill patients in whom *Candida* colonization was prospectively assessed, a "Candida score" of >2.5 (1 point for TPN, surgery, multifocal candida; 2 points for severe sepsis) accurately selected patients who would benefit from early antifungal treatment. The overall results from this study demonstrated a sensitivity of 81%, a specificity of 74%, a PPV of 16%, and an NPV of 98% [96].

The Ostrosky-Zeichner clinical prediction rule measures factors associated with the development of candidemia and IC. These factors include mechanical ventilation $\geq$ 48 h, systemic antibiotic use, and CVP (on any of days 1–3 of ICU admission), plus $\geq$ 1 of the following: any major surgery (days 7–0), pancreatitis (days 7–0), use of steroids/other immunosuppressive agents (days 7–0), use of TPN (days 1–3), or dialysis (days 1–3). The results revealed a sensitivity of 81%, a specificity of 74%, a PPV of 16%, and an NPV = 98% [97].

All these studies and the subsequent scores have limitations. These limitations include the potential for inaccurate predictions when trying to apply them to all critical patients, the reliance on clinical data that may not be readily available, the need for further validation in diverse settings, and their inability to definitively diagnose infection without confirmatory cultures. In essence, they should be used as tools to develop a risk assessment, not a diagnostic test [4,22,54].

Antifungal stewardship programs can be very valuable in initiating appropriate antifungal therapy. The combination of the "*Candida* Score" which was published by Posteraro et al. and the BDG assay can be a very useful tool to initiate early empiric therapy and discontinue antifungal therapy [97,98]. The criteria to initiate empiric or early presumptive antifungal therapy (echinocandin) in our institution include a *Candida* score of $\geq$3, high-risk patients, and a duration of >72 h on broad-spectrum antibiotics. If the patients meet these criteria, an antifungal is initiated after drawing a BDG assay. After 48 h, the BDG assay is repeated. If the BDG assay is positive, the antifungal is continued. However, if both BDG assays are negative, then the antifungal is discontinued. The use of this practice has helped decrease the use of empiric antifungal therapy by over 50% (Tables 7 and 8). A randomized multicenter study evaluating this methodology is still needed for it to be a strong recommendation.

Table 7. Antifungal stewardship guide for early presumptive therapy in high-risk patients with suspected Candidiasis: a stepwise approach [97,98].

• Patient on broad-spectrum antimicrobials for >72 h; • Negative blood cultures; • Candida score > 3.
Consider the following: • Obtaining β-D-glucan and one set of blood cultures; • Initiating echinocandin.
At 48–72 h, perform the following: • Repeat β-D-Glucan and one set of blood cultures.
If two sets of blood cultures are negative and two sets of β-D-glucan are negative, consider discontinuing the antifungal therapy.

Table 8. Candida score calculation [97,98].

Candida Score Components	Points Assigned
Total Parenteral Nutrition via Central Line	1
Surgical ICU Admission	1
Multifocal Candida Colonization	1
Sepsis	2

12. Conclusions

Infections due to *Candida* species are much more common than we think and remain a significant cause of morbidity and mortality in high-risk hosts [35]. These high rates are due to the lack of a diagnostic assay and the late initiation of appropriate antifungal therapy. Numerous challenges remain despite over 40 years of studying these infections, specifically, the highly variable and non-specific clinical presentations and the fact that no single, sensitive, universally applicable diagnostic assay is readily available. The current management of therapy needs to include having a high index of suspicion in high-risk groups, accurate knowledge of *Candida* species within each institution due to the increase in non-*albicans Candida* species, and the early initiation of appropriate antifungal therapy. Waiting for definitive proof of an infection increases the risk of a potentially fatal progression of the infection. Newer antifungal agents currently being studied may assist in the definitive therapy of these multidrug-resistant *Candida* species.

Funding: This research received no external funding.

Conflicts of Interest: J.A.V.: Consultant for Melinta, Basilea, Scynexis; L.W., A.Z. no conflict of interest.

References

1. World Health Organization (WHO) Report. WHO Fungal Priority Pathogens List to Guide Research, Development, and Public Health Action. Available online: https://www.who.int/publications/i/item/9789240060241 (accessed on 25 October 2022).
2. Parnums, D.V. Editorial: The World Health Organization (WHO) Fungal Priority Pathogens List in Response to Emerging Fungal Pathogens During the COVID-19 Pandemic. *Med. Sci. Monit.* **2022**, *28*, e939088. [CrossRef]
3. Centers for Disease Control and Prevention (CDC). Fungal Diseases and COVID-19. Available online: https://www.cdc.gov/fungal/covid-fungal.html (accessed on 2 March 2022).
4. McCarty, T.P.; White, C.M.; Pappas, P.G. Candidemia and Invasive Candidiasis. *Infect. Dis. Clin. N. Am.* **2021**, *35*, 389–413. [CrossRef]
5. Paphitou, N.I.; Ostrosky-Zeichner, L.; Rex, J.H. Rules for identifying patients at increased risk for candidal infections in the surgical intensive care unit: Approach to developing practical criteria for systematic use in antifungal prophylaxis trials. *Med. Mycol.* **2005**, *43*, 235–243. [CrossRef] [PubMed]
6. Kullberg, B.J.; Arendrup, M.C. Invasive Candidiasis. *N. Engl. J. Med.* **2015**, *373*, 1445–1456. [CrossRef] [PubMed]
7. Giacobbe, D.R.; Maraolo, A.E.; Simeon, V.; Magnè, F.; Pace, M.C.; Gentile, I.; Chiodini, P.; Viscoli, C.; Sanguinetti, M.; Mikulska, M.; et al. Changes in the relative prevalence of candidaemia due to non-*albicans Candida* species in adult in-patients: A systematic review, meta-analysis and meta-regression. *Mycoses* **2020**, *63*, 334–342. [CrossRef]
8. Forsberg, K.; Woodworth, K.; Walters, M.; Berkow, E.L.; Jackson, B.; Chiller, T.; Vallabhaneni, S. *Candida auris*: The recent emergence of a multidrug-resistant fungal pathogen. *Med. Mycol.* **2019**, *57*, 1–12, Erratum in *Med. Mycol.* **2019**, *57*, e7. [CrossRef] [PubMed]
9. Benedict, K.; Forsberg, K.; Gold, J.A.; Baggs, J.; Lyman, M. *Candida auris*—Associated Hospitalizations, United States, 2017–2022. *Emerg. Infect. Dis.* **2023**, *29*, 1485–1487. [CrossRef] [PubMed] [PubMed Central]
10. Lyman, M.; Forsberg, K.; Sexton, D.J.; Chow, N.A.; Lockhart, S.R.; Jackson, B.R.; Chiller, T. Worsening Spread of *Candida auris* in the United States, 2019 to 2021. *Ann. Intern. Med.* **2023**, *176*, 489–495. [CrossRef] [PubMed] [PubMed Central]
11. Vazquez, J.A.; Sanchez, V.; Dmuchowski, C.; Dembry, L.M.; Sobel, J.D.; Zervos, M.J. Nosocomial acquisition of *Candida albicans*: An epidemiologic study. *J. Infect. Dis.* **1993**, *168*, 195–201. [CrossRef] [PubMed]
12. Fidel, P.L., Jr.; Vazquez, J.A.; Sobel, J.D. *Candida glabrata*: Review of epidemiology, pathogenesis, and clinical disease with comparison to *C. albicans*. *Clin. Microbiol. Rev.* **1999**, *12*, 80–96. [CrossRef] [PubMed] [PubMed Central]
13. Cornely, O.A.; Bassetti, M.; Calandra, T.; Garbino, J.; Kullberg, B.J.; Lortholary, O.; Meersseman, W.; Akova, M.; Arendrup, M.C.; Arikan-Akdagli, S.; et al. ESCMID* guideline for the diagnosis and management of Candida diseases 2012: Non-neutropenic adult patients. *Clin. Microbiol. Infect.* **2012**, *18* (Suppl. S7), 19–37. [CrossRef] [PubMed]
14. Garnacho-Montero, J.; Díaz-Martín, A.; Cantón-Bulnes, L.; Ramírez, P.; Sierra, R.; Arias-Verdú, D.; Rodríguez-Delgado, M.; Loza-Vázquez, A.; Rodriguez-Gomez, J.; Gordón, M.; et al. Initial Antifungal Strategy Reduces Mortality in Critically Ill Patients With Candidemia: A Propensity Score–Adjusted Analysis of a Multicenter Study*. *Crit. Care Med.* **2018**, *46*, 384–393. [CrossRef] [PubMed]
15. De Pascale, G.; Martin-Loeches, I.; Nseir, S. Antifungal stewardship in critically ill patients. *Intensive Care Med.* **2023**, *49*, 681–684. [CrossRef] [PubMed] [PubMed Central]
16. De Pascale, G.; Posteraro, B.; D'arrigo, S.; Spinazzola, G.; Gaspari, R.; Bello, G.; Montini, L.M.; Cutuli, S.L.; Grieco, D.L.; Di Gravio, V.; et al. (1,3)-β-d-Glucan-based empirical antifungal interruption in suspected invasive candidiasis: A randomized trial. *Crit. Care* **2020**, *24*, 550. [CrossRef] [PubMed] [PubMed Central]
17. Cornely, O.; Pappas, P.G.; Krause, R.; Miller, R.; Ostrosky-Zeichner, L.; Vazquez, J.; Sanders, J.W.; Munoz, P.; Spec, A.; Thompson, G.R.; et al. Oral Ibrexafungerp Outcomes in Patients with Invasive Candidiasis and Candidemia from the FURI and CARES Studies. In Proceedings of the ECCMID, Lisbon, Portugal, 23–26 April 2022. Poster.
18. Mayer, F.L.; Wilson, D.; Hube, B. *Candida albicanspathogenicity* mechanisms. *Virulence* **2013**, *4*, 119–128. [CrossRef] [PubMed] [PubMed Central]
19. Mba, I.E.; Nweze, E.I. Mechanism of *Candida* pathogenesis: Revisiting the vital drivers. *Eur. J. Clin. Microbiol. Infect. Dis.* **2020**, *39*, 1797–1819. [CrossRef] [PubMed]
20. Soriano, A.; Honore, P.M.; Puerta-Alcalde, P.; Garcia-Vidal, C.; Pagotto, A.; Gonçalves-Bradley, D.C.; Verweij, P.E. Invasive candidiasis: Current clinical challenges and unmet needs in adult populations. *J. Antimicrob. Chemother.* **2023**, *78*, 1569–1585. [CrossRef] [PubMed] [PubMed Central]
21. Sagué, C.M.B.; Jarvis, W.R.; System, N.N.I.S. Secular trends in the epidemiology of nosocomial fungal infections in the United States, 1980-1990. National Nosocomial Infections Surveillance System. *J. Infect. Dis.* **1993**, *167*, 1247–1251. [CrossRef] [PubMed]

22. Pappas, P.G.; Lionakis, M.S.; [illegible] Zeichner, L.; Kullberg, B.J. Invasive Candidiasis. *Nat. Rev. Dis. Primers* **2018**, *4*, 18026. [CrossRef] [PubMed]
23. Tsay, S.V.; Mu, Y.; Williams, S.; Epson, E.; Nadle, J.; Bamberg, W.M.; Barter, D.M.; Johnston, H.L.; Farley, M.M.; Harb, S.; et al. Burden of Candidemia in the United States, 2017. *Clin. Infect. Dis.* **2020**, *71*, e449–e453. [CrossRef] [PubMed]
24. Bassetti, M.; Giacobbe, D.R.; Vena, A.; Trucchi, C.; Ansaldi, F.; Antonelli, M.; Adamkova, V.; Alicino, C.; Almyroudi, M.-P.; Atchade, E.; et al. Incidence and outcome of invasive candidiasis in intensive care units (ICUs) in Europe: Results of the EUCANDICU project. *Crit. Care* **2019**, *23*, 219. [CrossRef] [PubMed] [PubMed Central]
25. Bassetti, M.; Giacobbe, D.R.; Agvald-Ohman, C.; Akova, M.; Alastruey-Izquierdo, A.; Arikan-Akdagli, S.; Azoulay, E.; Blot, S.; Cornely, O.A.; Cuenca-Estrella, M.; et al. Invasive Fungal Diseases in Adult Patients in Intensive Care Unit (FUNDICU): 2024 consensus definitions from ESGCIP, EFISG, ESICM, ECMM, MSGERC, ISAC, and ISHAM. *Intensive Care Med.* **2024**, *50*, 502–515. [CrossRef] [PubMed] [PubMed Central]
26. Vincent, J.-L.; Sakr, Y.; Singer, M.; Martin-Loeches, I.; Machado, F.R.; Marshall, J.C.; Finfer, S.; Pelosi, P.; Brazzi, L.; Aditianingsih, D.; et al. Prevalence and Outcomes of Infection Among Patients in Intensive Care Units in 2017. *JAMA* **2020**, *323*, 1478–1487. [CrossRef] [PubMed] [PubMed Central]
27. Thomas-Rüddel, D.O.; Schlattmann, P.; Pletz, M.; Kurzai, O.; Bloos, F. Risk Factors for Invasive Candida Infection in Critically Ill Patients: A Systematic Review and Meta-analysis. *Chest* **2022**, *161*, 345–355. [CrossRef] [PubMed] [PubMed Central]
28. Giamarellou, H.; Antoniadou, A. Epidemiology, diagnosis, and therapy of fungal infections in surgery. *Infect. Control Hosp. Epidemiol.* **1996**, *17*, 558–564. [CrossRef] [PubMed]
29. Habighorst, K.; Sanders, J.M.; Hennessy, S.A.; Goff, K.; Wan, B.; Johns, M. Identification of Risk Factors for Intra-Abdominal Candidiasis. *Surg. Infect.* **2023**, *24*, 910–915. [CrossRef] [PubMed]
30. Rex, J.H. Treatment of candidemia in the non-neutropenic host. In Proceedings of the 14th Congress of the International Society for Human and Animal Mycology, Buenos Aires, Argentina, 8–12 May 2000; Speaker in symposium entitled "Superficial and invasive infections caused by *Candida* and other emerging yeasts.".
31. Ostrosky-Zeichner, L. New approaches to the risk of Candida in the intensive care unit. *Curr. Opin. Infect. Dis.* **2003**, *16*, 533–537. [CrossRef] [PubMed]
32. Clancy, C.J.; Nguyen, M.H. Diagnosing Invasive Candidiasis. *J. Clin. Microbiol.* **2018**, *56*, e01909-17. [CrossRef] [PubMed] [PubMed Central]
33. Clancy, C.J.; Nguyen, M.H. Finding the "Missing 50%" of invasive candidiasis: How nonculture diagnostics will improve understanding of disease spectrum and transform patient care. *Clin. Infect. Dis.* **2013**, *56*, 1284–1292. [CrossRef] [PubMed]
34. Morrell, M.; Fraser, V.J.; Kollef, M.H. Delaying the Empiric Treatment of *Candida* Bloodstream Infection until Positive Blood Culture Results Are Obtained: A Potential Risk Factor for Hospital Mortality. *Antimicrob. Agents Chemother.* **2005**, *49*, 3640–3645. [CrossRef]
35. Roosen, J.; Frans, E.; Wilmer, A.; Knockaert, D.C.; Bobbaers, H. Comparison of Premortem Clinical Diagnoses in Critically Ill Patients and Subsequent Autopsy Findings. *Mayo Clin. Proc.* **2000**, *75*, 562–567. [CrossRef] [PubMed]
36. Logan, C.; Martin-Loeches, I.; Bicanic, T. Invasive candidiasis in critical care: Challenges and future directions. *Intensive Care Med.* **2020**, *46*, 2001–2014. [CrossRef] [PubMed]
37. Tran, T.; Beal, S.G. Application of the 1,3-β-d-Glucan (Fungitell) Assay in the Diagnosis of Invasive Fungal Infections. *Arch. Pathol. Lab. Med.* **2016**, *140*, 181–185. [CrossRef] [PubMed]
38. Maertens, J.; Miceli, M.H. Role of Non-Culture-Based Tests, with an Emphasis on Galactomannan Testing for the Diagnosis of Invasive Aspergillosis. *Semin. Respir. Crit. Care Med.* **2015**, *36*, 650–661. [CrossRef] [PubMed]
39. McKeating, C.; White, P.L.; Posso, R.; Palmer, M.; Johnson, E.; McMullan, R. Diagnostic accuracy of fungal PCR and β-d-glucan for detection of candidaemia: A preliminary evaluation. *J. Clin. Pathol.* **2018**, *71*, 420–424. [CrossRef] [PubMed]
40. Jaijakul, S.; Vazquez, J.A.; Swanson, R.N.; Ostrosky-Zeichner, L. (1,3)-D-Glucan as a prognostic marker of treatment response in invasive candidiasis. *Clin. Infect. Dis.* **2012**, *55*, 521–526. [CrossRef] [PubMed]
41. Mylonakis, E.; Clancy, C.J.; Ostrosky-Zeichner, L.; Garey, K.W.; Alangaden, G.J.; Vazquez, J.A.; Groeger, J.S.; Judson, M.A.; Vinagre, Y.-M.; Heard, S.O.; et al. T2 magnetic resonance assay for the rapid diagnosis of candidemia in whole blood: A clinical trial. *Clin. Infect. Dis.* **2015**, *60*, 892–899. [CrossRef] [PubMed]
42. Clancy, C.J.; Nguyen, M.H. Non-Culture Diagnostics for Invasive Candidiasis: Promise and Unintended Consequences. *J. Fungi* **2018**, *4*, 27. [CrossRef] [PubMed] [PubMed Central]
43. Egger, M.; Horvath, A.; Prüller, F.; Fickert, P.; Finkelman, M.; Kriegl, L.; Grønbæk, H.; Møller, H.J.; Prattes, J.; Krause, R.; et al. Fungal translocation measured by serum 1,3-ß-D-glucan correlates with severity and outcome of liver cirrhosis—A pilot study. *Liver Int.* **2023**, *43*, 1975–1983. [CrossRef] [PubMed] [PubMed Central]

44. Hoenigl, M.; Pérez-Santiago, J.; Nakazawa, M.; de Oliveira, M.F.; Zhang, Y.; Finkelman, M.A.; Letendre, S.; Smith, D.; Gianella, S. (1→3)-β-d-Glucan: A Biomarker for Microbial Translocation in Individuals With Acute or Early HIV Infection? *Front. Immunol.* **2016**, *7*, 404. [CrossRef] [PubMed] [PubMed Central]
45. Zacharioudakis, I.M.; Zervou, F.N.; Mylonakis, E. Use of T2MR in invasive candidiasis with and without candidemia. *Futur. Microbiol.* **2018**, *13*, 1165–1173. [CrossRef] [PubMed]
46. Tang, D.-L.; Chen, X.; Zhu, C.-G.; Li, Z.-W.; Xia, Y.; Guo, X.-G. Pooled analysis of T2 Candida for rapid diagnosis of candidiasis. *BMC Infect. Dis.* **2019**, *19*, 798. [CrossRef] [PubMed] [PubMed Central]
47. Pfaller, M.A.; Castanheira, M. Nosocomial Candidiasis: Antifungal Stewardship and the Importance of Rapid Diagnosis. *Med. Mycol.* **2016**, *54*, 1–22. [CrossRef] [PubMed]
48. Calandra, T.; Roberts, J.A.; Antonelli, M.; Bassetti, M.; Vincent, J.-L. Diagnosis and management of invasive candidiasis in the ICU: An updated approach to an old enemy. *Crit. Care* **2016**, *20*, 125. [CrossRef] [PubMed] [PubMed Central]
49. Camp, I.; Spettel, K.; Willinger, B. Molecular Methods for the Diagnosis of Invasive Candidiasis. *J. Fungi* **2020**, *6*, 101. [CrossRef] [PubMed] [PubMed Central]
50. Senchyna, F.; A Hogan, C.; Murugesan, K.; Moreno, A.; Ho, D.Y.; Subramanian, A.; Schwenk, H.T.; Budvytiene, I.; A Costa, H.; Gombar, S.; et al. Clinical Accuracy and Impact of Plasma Cell-Free DNA Fungal Polymerase Chain Reaction Panel for Noninvasive Diagnosis of Fungal Infection. *Clin. Infect. Dis.* **2021**, *73*, 1677–1684. [CrossRef] [PubMed]
51. Musher, D.M. Problems with etiologic diagnosis of community-acquired pneumonia using plasma microbial cell-free DNA sequencing. *Antimicrob. Steward. Healthc. Epidemiol.* **2023**, *3*, e206. [CrossRef] [PubMed] [PubMed Central]
52. Gu, W.; Deng, X.; Lee, M.; Sucu, Y.D.; Arevalo, S.; Stryke, D.; Federman, S.; Gopez, A.; Reyes, K.; Zorn, K.; et al. Rapid pathogen detection by metagenomic next-generation sequencing of infected body fluids. *Nat. Med.* **2021**, *27*, 115–124. [CrossRef] [PubMed] [PubMed Central]
53. Gu, W.; Miller, S.; Chiu, C.Y. Clinical Metagenomic Next-Generation Sequencing for Pathogen Detection. *Annu. Rev. Pathol. Mech. Dis.* **2019**, *14*, 319–338. [CrossRef] [PubMed] [PubMed Central]
54. Pappas, P.G.; Kauffman, C.A.; Andes, D.R.; Clancy, C.J.; Marr, K.A.; Ostrosky-Zeichner, L.; Reboli, A.C.; Schuster, M.G.; Vazquez, J.A.; Walsh, T.J.; et al. Clinical Practice Guideline for the Management of Candidiasis: 2016 Update by the Infectious Diseases Society of America. *Clin. Infect. Dis.* **2016**, *62*, e1–e50. [CrossRef] [PubMed] [PubMed Central]
55. Lehman, A.; Tessier, K.M.; Sattarova, V.; Montezuma, S.R.; Kline, S.; Erayil, S.E. Do Patients With Candidemia Need an Ophthalmologic Examination? *Open Forum Infect. Dis.* **2024**, *11*, ofae663. [CrossRef] [PubMed] [PubMed Central]
56. Zilberberg, M.D.; Kollef, M.H.; Arnold, H.; Labelle, A.; Micek, S.T.; Kothari, S.; Shorr, A.F. Inappropriate empiric antifungal therapy for candidemia in the ICU and hospital resource utilization: A retrospective cohort study. *BMC Infect. Dis.* **2010**, *10*, 150. [CrossRef] [PubMed]
57. Kollef, M.; Micek, S.; Hampton, N.; Doherty, J.A.; Kumar, A. Septic Shock Attributed to Candida Infection: Importance of Empiric Therapy and Source Control. *Clin. Infect. Dis.* **2012**, *54*, 1739–1746. [CrossRef]
58. Timsit, J.-F.; Azoulay, E.; Schwebel, C.; Charles, P.E.; Cornet, M.; Souweine, B.; Klouche, K.; Jaber, S.; Trouillet, J.-L.; Bruneel, F.; et al. Empirical Micafungin Treatment and Survival Without Invasive Fungal Infection in Adults With ICU-Acquired Sepsis, *Candida* Colonization, and Multiple Organ Failure: The EMPIRICUS Randomized Clinical Trial. *JAMA* **2016**, *316*, 1555–1564. [CrossRef]
59. Ostrosky-Zeichner, L.; Shoham, S.; Vazquez, J.; Reboli, A.; Betts, R.; Barron, M.A.; Schuster, M.; Judson, M.A.; Revankar, S.G.; Caeiro, J.P.; et al. MSG-01: A randomized, double-blind, placebo-controlled trial of caspofungin prophylaxis followed by preemptive therapy for invasive candidiasis in high-risk adults in the critical care setting. *Clin. Infect. Dis.* **2014**, *58*, 1219–1226. [CrossRef]
60. Knitsch, W.; Vincent, J.-L.; Utzolino, S.; François, B.; Dinya, T.; Dimopoulos, G.; Özgüneş, I.; Valía, J.C.; Eggimann, P.; León, C.; et al. A randomized, placebo-controlled trial of preemptive antifungal therapy for the prevention of invasive candidiasis following gastrointestinal surgery for intra-abdominal infections. *Clin. Infect. Dis.* **2015**, *61*, 1671–1678. [CrossRef] [PubMed] [PubMed Central]
61. Schuster, M.G.; Edwards, J.E.; Sobel, J.D.; Darouiche, R.O.; Karchmer, A.W.; Hadley, S.; Slotman, G.; Panzer, H.; Biswas, P.; Rex, J.H. Empirical fluconazole versus placebo for intensive care unit patients: A randomized trial. *Ann. Intern. Med.* **2008**, *149*, 83–90. [CrossRef] [PubMed]
62. Ferreira, D.; Grenouillet, F.; Blasco, G.; Samain, E.; Hénon, T.; Dussaucy, A.; Millon, L.; Mercier, M.; Pili-Floury, S. Outcomes associated with routine systemic antifungal therapy in critically ill patients with Candida colonization. *Intensive Care Med.* **2015**, *41*, 1077–1088. [CrossRef] [PubMed]

63. Fournier, P.; Schwebel, C.; [illegible] Foroni, L.; Hamidfar-Roy, R.; Cornet, M.; Timsit, J.-F.; Pelloux, H. Antifungal use influences Candida species distribution and susceptibility in the intensive care unit. *J. [illegible]* 66, 2880–2886. [CrossRef] [PubMed]
64. Thomaz, D.Y.; de Almeida, J.N.; Sejas, O.N.E.; Del Negro, G.M.B.; Carvalho, G.O.M.H.; Gimenes, V.M.F.; de Souza, M.; Arastehfar, A.; Camargo, C.; Motta, A.; et al. Environmental Clonal Spread of Azole-Resistant *Candida parapsilosis* with Erg11-Y132F Mutation Causing a Large Candidemia Outbreak in a Brazilian Cancer Referral Center. *J. Fungi* **2021**, *7*, 259. [CrossRef] [PubMed] [PubMed Central]
65. Arastehfar, A.; Hilmioğlu-Polat, S.; Daneshnia, F.; Pan, W.; Hafez, A.; Fang, W.; Liao, W.; Şahbudak-Bal, Z.; Metin, D.Y.; De Almeida, J.N., Jr.; et al. Clonal Candidemia Outbreak by Candida parapsilosis Carrying Y132F in Turkey: Evolution of a Persisting Challenge. *Front. Cell. Infect. Microbiol.* **2021**, *11*, 676177. [CrossRef] [PubMed] [PubMed Central]
66. Nivoix, Y.; Launoy, A.; Lutun, P.; Moulin, J.-C.; Phai Pang, K.-A.; Fornecker, L.-M.; Wolf, M.; Levêque, D.; Letscher-Bru, V.; Beretz, L.; et al. Adherence to recommendations for the use of antifungal agents in a tertiary care hospital. *J. Antimicrob. Chemother.* **2012**, *67*, 2506–2513. [CrossRef] [PubMed]
67. Jacobs, D.M.; Dilworth, T.J.; Beyda, N.D.; Casapao, A.M.; Bowers, D.R. Overtreatment of Asymptomatic Candiduria among Hospitalized Patients: A Multi-institutional Study. *Antimicrob. Agents Chemother.* **2017**, *62*, e01464-17. [CrossRef] [PubMed] [PubMed Central]
68. Johnson, M.D.; Lewis, R.E.; Ashley, E.S.D.; Ostrosky-Zeichner, L.; Zaoutis, T.; Thompson, G.R.; Andes, D.R.; Walsh, T.J.; Pappas, P.G.; A Cornely, O.; et al. Core Recommendations for Antifungal Stewardship: A Statement of the Mycoses Study Group Education and Research Consortium. *J. Infect. Dis.* **2020**, *222* (Suppl. S3), S175–S198. [CrossRef] [PubMed] [PubMed Central]
69. Bartalucci, C.; Giacobbe, D.R.; Vena, A.; Bassetti, M. Empirical Therapy for Invasive Candidiasis in Critically Ill Patients. *Curr. Fungal Infect. Rep.* **2024**, *18*, 136–145. [CrossRef]
70. Patch, M.E.; Weisz, E.; Cubillos, A.; Estrada, S.J.; A Pfaller, M. Impact of rapid, culture-independent diagnosis of candidaemia and invasive candidiasis in a community health system. *J. Antimicrob. Chemother.* **2018**, *73* (Suppl. S4), iv27–iv30. [CrossRef] [PubMed]
71. O'donnell, M.; Shields, R.K.; Marini, R.V.; Groetzinger, L.M.; Potoski, B.A.; Falcione, B.A.; Shah, S.; McCreary, E.K.; Clarke, L.; Brant, E.; et al. Stewardship-Guided T2Candida Testing Shortens Time to Antifungal Treatment and Reduces Antifungal Usage Among Medical Intensive Care Unit Patients With Septic Shock. *Open Forum Infect. Dis.* **2023**, *10*, ofad538. [CrossRef] [PubMed] [PubMed Central]
72. Helweg-Larsen, J.; Steensen, M.; Pedersen, F.M.; Jensen, P.B.; Perch, M.; Møller, K.; Olesen, B.R.; Søderlund, M.; Arendrup, M.C. Intensive Care Antifungal Stewardship Programme Based on T2Candida PCR and *Candida* Mannan Antigen: A Prospective Study. *J. Fungi* **2021**, *7*, 1044. [CrossRef]
73. Bienvenu, A.-L.; Pradat, P.; Guerin, C.; Aubrun, F.; Fellahi, J.-L.; Friggeri, A.; Guichon, C.; Hernu, R.; Menotti, J.; Monard, C.; et al. Evaluation of first-line therapies for the treatment of candidemia in ICU patients: A propensity score analysis. *Int. J. Infect. Dis.* **2020**, *93*, 15–21. [CrossRef] [PubMed]
74. Martin-Loeches, I.; Antonelli, M.; Cuenca-Estrella, M.; Dimopoulos, G.; Einav, S.; De Waele, J.J.; Garnacho-Montero, J.; Kanj, S.S.; Machado, F.R.; Montravers, P.; et al. ESICM/ESCMID task force on practical management of invasive candidiasis in critically ill patients. *Intensive Care Med.* **2019**, *45*, 789–805. [CrossRef] [PubMed]
75. Szymański, M.; Chmielewska, S.; Czyżewska, U.; Malinowska, M.; Tylicki, A. Echinocandins–structure, mechanism of action and use in antifungal therapy. In *Journal of Enzyme Inhibition and Medicinal Chemistry*; Taylor and Francis Ltd.: London, UK, 2022; pp. 876–894.
76. Tsay, S.; Kallen, A.; Jackson, B.R.; Chiller, T.M.; Vallabhaneni, S. Approach to the Investigation and Management of Patients With Candida auris, an Emerging Multidrug-Resistant Yeast. *Clin. Infect. Dis.* **2018**, *66*, 306–311. [CrossRef] [PubMed] [PubMed Central]
77. Tsay, S.; Welsh, R.M.; Adams, E.H.; Chow, N.A.; Gade, L.; Tsay, S.; Welsh, R.M.; Adams, E.H.; Chow, N.A.; Gade, L.; et al. Notes from the Field: Ongoing Transmission of *Candida auris* in Health Care Facilities—United States, June 2016–May 2017. *Mmwr-Morbidity Mortal. Wkly. Rep.* **2017**, *66*, 514–515. [CrossRef] [PubMed] [PubMed Central]
78. Prasad, R.; Shah, A.H.; Rawal, M.K. Antifungals: Mechanism of action and drug resistance. *Adv. Exp. Med. Biol.* **2016**, *892*, 327–349.
79. Cavalheiro, M.; Teixeira, M.C. Candida Biofilms: Threats, Challenges, and Promising Strategies. *Front. Med.* **2018**, *5*, 28. [CrossRef] [PubMed] [PubMed Central]
80. Vazquez, J.; Reboli, A.C.; Pappas, P.G.; Patterson, T.F.; Reinhardt, J.; Chin-Hong, P.; Tobin, E.; Kett, D.H.; Biswas, P.; Swanson, R. Evaluation of an early step-down strategy from intravenous anidulafungin to oral azole therapy for the treatment of candidemia and other forms of invasive candidiasis: Results from an open-label trial. *BMC Infect. Dis.* **2014**, *14*, 97. [CrossRef] [PubMed] [PubMed Central]
81. Bailly, S.; Leroy, O.; Montravers, P.; Constantin, J.-M.; Dupont, H.; Guillemot, D.; Lortholary, O.; Mira, J.-P.; Perrigault, P.-F.; Gangneux, J.-P.; et al. Antifungal de-escalation was not associated with adverse outcome in critically ill patients treated for invasive candidiasis: Post hoc analyses of the AmarCAND2 study data. *Intensive Care Med.* **2015**, *41*, 1931–1940. [CrossRef] [PubMed]

82. Moreno-García, E.; Puerta-Alcalde, P.; Gariup, G.; Fernández-Ruiz, M.; Cortés, L.E.L.; Cuervo, G.; Salavert, M.; Merino, P.; Machado, M.; Guinea, J.; et al. Early Stepdown From Echinocandin to Fluconazole Treatment in Candidemia: A Post Hoc Analysis of Three Cohort Studies. *Open Forum Infect. Dis.* **2021**, *8*, ofab250. [CrossRef] [PubMed] [PubMed Central]
83. Jaffal, K.; Poissy, J.; Rouze, A.; Preau, S.; Sendid, B.; Cornu, M.; Nseir, S. De-escalation of antifungal treatment in critically ill patients with suspected invasive Candida infection: Incidence, associated factors, and safety. *Ann. Intensive Care* **2018**, *8*, 49. [CrossRef] [PubMed] [PubMed Central]
84. Rouzé, A.; Loridant, S.; Poissy, J.; Dervaux, B.; Sendid, B.; Cornu, M.; Nseir, S.; S-TAFE Study Group. Biomarker-based strategy for early discontinuation of empirical antifungal treatment in critically ill patients: A randomized controlled trial. *Intensive Care Med.* **2017**, *43*, 1668–1677. [CrossRef] [PubMed]
85. Bassetti, M.; Righi, E.; Ansaldi, F.; Merelli, M.; Cecilia, T.; De Pascale, G.; Diaz-Martin, A.; Luzzati, R.; Rosin, C.; Lagunes, L.; et al. A multicenter study of septic shock due to candidemia: Outcomes and predictors of mortality. *Intensive Care Med.* **2014**, *40*, 839–845. [CrossRef] [PubMed]
86. Colombo, R.E.; Vazquez, J.A. An evaluation of ibrexafungerp for the treatment of invasive candidiasis: The evidence to date. *Expert Opin. Pharmacother.* **2021**, *22*, 797–807. [CrossRef] [PubMed]
87. August, B.A.; Kale-Pradhan, P.B. Management of invasive candidiasis: A focus on rezafungin, ibrexafungerp, and fosmanogepix. *Pharmacother. J. Hum. Pharmacol. Drug Ther.* **2024**, *44*, 467–479. [CrossRef] [PubMed]
88. Vazquez, J.A.; Pappas, P.G.; Boffard, K.; Paruk, F.; Bien, P.A.; Tawadrous, M.; Ople, E.; Wedel, P.; Oborska, I.; Hodges, M.R. Clinical Efficacy and Safety of a Novel Antifungal, Fosmanogepix, in Patients with Candidemia Caused by Candida auris : Results from a Phase 2 Trial. *Antimicrob. Agents Chemother.* **2023**, *67*, e0141922. [CrossRef] [PubMed] [PubMed Central]
89. Pappas, P.G.; A Vazquez, J.; Oren, I.; Rahav, G.; Aoun, M.; Bulpa, P.; Ben-Ami, R.; Ferrer, R.; Mccarty, T.; Thompson, G.R.; et al. Clinical safety and efficacy of novel antifungal, fosmanogepix, for the treatment of candidaemia: Results from a Phase 2 trial. *J. Antimicrob. Chemother.* **2023**, *78*, 2471–2480. [CrossRef] [PubMed] [PubMed Central]
90. Bader, J.C.; Lakota, E.A.; Flanagan, S.; Ong, V.; Sandison, T.; Rubino, C.M.; Bhavnani, S.M.; Ambrose, P.G. Overcoming the resistance hurdle: Pharmacokinetic-pharmacodynamic target attainment analyses for rezafungin (CD101) against *Candida albicans* and *Candida glabrata*. *Antimicrob. Agents Chemother.* **2018**, *62*, e02614-17. [CrossRef] [PubMed]
91. Sharma, D.; Vazquez, J.A. An evaluation of rezafungin: The latest treatment option for adults with invasive candidiasis An Evaluation of Rezafungin: The latest treatment option for adults with candidemia and invasive candidiasis. *Expert Opin. Pharmacother.* **2024**, *25*, 339–347.
92. Thompson, G.R.; Soriano, A.; Skoutelis, A.; A Vazquez, J.; Honore, P.M.; Horcajada, J.P.; Spapen, H.; Bassetti, M.; Ostrosky-Zeichner, L.; Das, A.F.; et al. Rezafungin Versus Caspofungin in a Phase 2, Randomized, Double-blind Study for the Treatment of Candidemia and Invasive Candidiasis: The STRIVE Trial. *Clin. Infect. Dis.* **2021**, *73*, e3647–e3655, Erratum in *Clin. Infect. Dis.* **2021**, *73*, 561–562. [CrossRef] [PubMed] [PubMed Central]
93. Thompson, G.R.; Soriano, A.; A Cornely, O.; Kullberg, B.J.; Kollef, M.; Vazquez, J.; Honore, P.M.; Bassetti, M.; Pullman, J.; Chayakulkeeree, M.; et al. Rezafungin versus caspofungin for treatment of candidaemia and invasive candidiasis (ReSTORE): A multicentre, double-blind, double-dummy, randomised phase 3 trial. *Lancet* **2023**, *401*, 49–59. [CrossRef] [PubMed]
94. Eggimann, P.; Pittet, D. Candida colonization index and subsequent infection in critically ill surgical patients: 20 years later. *Intensive Care Med.* **2014**, *40*, 1429–1448. [CrossRef] [PubMed] [PubMed Central]
95. Pittet, D.; Monod, M.; Suter, P.M.; Frenk, E.; Auckenthaler, R. *Candida* colonization and subsequent infections in critically ill surgical patients. *Ann. Surg.* **1994**, *220*, 751–758. [CrossRef] [PubMed] [PubMed Central]
96. Posteraro, B.; De Pascale, G.; Tumbarello, M.; Torelli, R.; Pennisi, M.A.; Bello, G.; Maviglia, R.; Fadda, G.; Sanguinetti, M.; Antonelli, M. Early diagnosis of candidemia in intensive care unit patients with sepsis: A prospective comparison of (1→3)-β-D-glucan assay, Candida score, and colonization index. *Crit. Care* **2011**, *15*, R249. [CrossRef] [PubMed] [PubMed Central]
97. Ostrosky-Zeichner, L.; Pappas, P.G.; Shoham, S.; Reboli, A.; Barron, M.A.; Sims, C.; Wood, C.; Sobel, J.D. Improvement of a clinical prediction rule for clinical trials on prophylaxis for invasive candidiasis in the intensive care unit. *Mycoses* **2011**, *54*, 46–51. [CrossRef] [PubMed]
98. Eggimann, P.; Marchetti, O. Is (1→3)-β-D-glucan the missing link from bedside assessment to pre-emptive therapy of invasive candidiasis? *Crit. Care* **2011**, *15*, 1017. [CrossRef] [PubMed] [PubMed Central]

Journal of *Fungi*

MDPI

Article

Epidemiology and Inpatient Outcomes of Invasive Aspergillosis in Patients with Liver Failure and Cirrhosis

Aditya Sharma [1], Bashar Mohamad [2] and Ayman O. Soubani [3,*]

1 Department of Internal Medicine, Wayne State University School of Medicine, Detroit, MI 48201, USA; adityamedicine@wayne.edu
2 Division of Gastroenterology, Wayne State University School of Medicine, Detroit, MI 48201, USA; gf2633@wayne.edu
3 Division of Pulmonary, Critical Care and Sleep Medicine, Wayne State University School of Medicine, Detroit, MI 48201, USA
* Correspondence: asoubani@med.wayne.edu; Tel.: +1-313-966-0695; Fax: +1-313-993-0562

Abstract: Objective: The aim of this study was to estimate the incidence and inpatient outcomes of liver failure and cirrhosis (LFC) admissions with invasive aspergillosis (IA) in the United States. Methods: This retrospective cohort study utilized the 2016–2020 National Inpatient Sample (NIS) database to analyze outcomes of IA in LFC admissions. Baseline variables, including demographics, comorbidities, and complications, were identified using International Classification of Diseases, Tenth Revision, Clinical Modification (ICD-10-CM) codes, and liver transplant admissions were excluded. Outcomes were compared between LFC admissions with and without IA. Results: During the study period, 9515 (0.36%) LFC admissions were associated with IA. This cohort experienced significantly higher rates of complications, including acute kidney injury (AKI) (73.36% vs. 42.96%; $p < 0.001$) and acute respiratory failure (ARF) (65.74% vs. 24.85%; $p < 0.001$). IA admissions required invasive mechanical ventilation (IMV) more frequently (58.17% vs. 18.78%; $p < 0.001$). All-cause inpatient mortality was significantly higher in the aspergillosis group (43.40% vs. 15.75%; $p < 0.001$). IA admissions had longer lengths of stay (LOS), with 38.89% exceeding 21 days compared to 6.20% ($p < 0.001$), and a mean LOS more than three times longer (22.9 vs. 7.5 days; $p < 0.001$). The IA group incurred over four times higher hospital charges (USD 459,414.9 vs. USD 104,389.4; $p < 0.001$) and hospitalization costs (USD 108,030.6 vs. USD 24,272.1; $p < 0.001$) compared to the LFC without aspergillosis group. Interpretation: LFC admissions with IA experienced poorer outcomes, longer hospital stays, and significantly higher healthcare costs, underscoring the need for targeted interventions in this high-risk, nonclassical population.

Keywords: invasive aspergillosis; liver failure; cirrhosis; fungal infection; invasive mechanical ventilation

1. Introduction

Aspergillosis is a rare but fatal non-traditional infection in patients with liver failure, with previous studies reporting mortality rates exceeding 50% [1–3]. In recent years, there has been an increasing trend in aspergillosis infections in the United States [4]. It typically begins with inhalation of *Aspergillus* spores into the lungs, followed by hematogenous spread to other organ systems. Traditionally, fungal infections were primarily associated with severely immunocompromised individuals, such as those with hematologic malignancies, neutropenia, or T-cell suppression [5]. However, these infections are now increasingly observed in other patient populations, including those with cirrhosis [2]. Notably, approximately 50% of patients with acute or advanced liver disease are diagnosed with invasive aspergillosis postmortem [1].

Despite these alarming statistics, there remains a significant lack of data on the outcomes of LFC patients with IA. This study aims to fill this gap by providing a comprehensive overview of the epidemiology, clinical outcomes, and socioeconomic impact of IA in LFC admissions, using the NIS database from 2016 to 2020. Leveraging this extensive administrative database enables the study of rare, low-incidence conditions like IA and facilitates the calculation of nationally representative estimates.

2. Methods

2.1. Data Source

This retrospective cohort study aimed to investigate the factors associated with LFC admissions with IA and their in-hospital mortality. Additionally, the study examined the relationship between LFC admissions and other socioeconomic and hospital-level factors, such as length of stay and hospitalization cost. The study utilized the NIS database for the years 2016 to 2020, developed for the Healthcare Cost and Utilization Project, sponsored by the Agency for Healthcare Research and Quality. The NIS is a publicly available, all-payer database of inpatient hospitalizations in the United States, providing nationally representative estimates of hospital inpatient stays. It contains a stratified sample of approximately 20% of all discharges from community hospitals in the country, estimating more than 35 million hospitalizations annually. Each observation in the database represents a hospitalization with one primary diagnosis, up to 39 secondary diagnoses, and 25 procedure diagnoses [6]. These are coded using the International Classification of Diseases, Tenth Revision, Clinical Modification (ICD-10-CM) codes and the International Classification of Diseases, Tenth Revision, Procedure Coding System (ICD-10-PCS) codes. The NIS data are depersonalized, and individual identities are protected. As the study used the NIS database, it was exempt from review by the institutional review board.

2.2. Study Population

We analyzed the 2016–2020 NIS database to identify LFC hospitalizations using ICD-10-CM codes for acute, alcoholic, and chronic liver failure (K7200, K7201, K7040, K7041, K7210, K7211, K7290, K7291, K717, K7030, K7031, K7460, K7469). Cirrhosis was defined by combining codes for cirrhosis and complications such as portal hypertension (K766), ascites (R188), spontaneous bacterial peritonitis (K652), hepatorenal syndrome (K767), hepatic encephalopathy (K7291, K7201, K7211, K7290), and esophageal varices (I8501, I8511, I8510, I8500) [7]. Liver transplant admissions were excluded from the study and identified using ICD-10-CM (T8640, T8641, T8642, T8643, T8649, Z4823, Z944) and ICD-10-PCS (0FY00Z0, 0FY00Z1, 0FY00Z2) codes. Aspergillosis was identified using ICD-10-CM codes B440, B441, B442, B447, B448, B4481, B4489, B449, B488, and B49 listed anywhere in the diagnosis column.

2.3. Baseline Variables and Comparison Groups

The study evaluated baseline variables, including demographics such as age, sex, median household income, and race, categorized as White, Black, Hispanic, and Other (encompassing Asian or Pacific Islanders and Native Americans). Comorbidities, including alcohol use disorder, cerebrovascular disease, coronary artery disease, drug use disorder,

of pulmonary comorbidities, including bronchiectasis (2.26% vs. 0.18%; $p < 0.001$) and cystic fibrosis (1.05% vs. 0.07%; $p < 0.001$). The IA group experienced a significantly higher incidence of complications compared to the non-aspergillosis LFC group. These included AKI, ARF, non-gastrointestinal bleeding, disseminated intravascular coagulation, intracranial hemorrhage, sepsis, and venous thromboembolism (Table 1).

Table 1. Baseline Variables of Admissions with Liver Failure and Cirrhosis With and Without Aspergillosis.

Variable	LFC Without Aspergillosis (N: 3,005,849)	LFC with Aspergillosis (N: 9515)	*p* Value
Age at admission in years, mean (±SD)	60.3 (13.5)	58.4 (14.3)	<0.001
Age group (in years):			<0.001
18–40	8.60%	12.24%	
41–60	40.46%	39.31%	
61–80	43.91%	43.41%	
>80	7.03%	5.04%	
Race:			0.0001
White	66.48%	61.66%	
Black	10.97%	13.58%	
Hispanic	15.47%	16.27%	
Others [a]	7.08%	8.49%	
Female	41.48%	43.09%	0.1632
Quarter:			0.1219
Quarter 1 (January–March)	24.93%	25.43%	
Quarter 2 (April–June)	24.36%	24.12%	
Quarter 3 (July–September)	25.11%	23.06%	
Quarter 4 (October–December)	25.61%	27.38%	
Median household income:			0.0036
Quartile 1 (poorest)	33.07%	30.77%	
Quartile 2	26.93%	26.14%	
Quartile 3	23.07%	23.02%	
Quartile 4 (wealthiest)	16.93%	20.06%	
Primary expected payer:			<0.001
Medicare	47.51%	45.22%	
Medicaid	23.31%	22.98%	
Private Insurance	20.45%	24.87%	
Self-pay and other [b]	8.73%	6.94%	
Admitted to teaching hospital	73.36%	83.39%	<0.001
Hospital bed size:			<0.001
Small	17.89%	12.45%	
Medium	28.00%	23.33%	
Large	54.11%	64.21%	

Table 1. Cont.

Variable	LFC Without Aspergillosis (N: 3,005,849)	LFC with Aspergillosis (N: 9515)	p Value
Region of Hospital:			0.0565
Northeast	16.42%	17.66%	
Midwest	20.58%	19.39%	
South	39.77%	37.52%	
West	23.22%	25.43%	
Charlson comorbidity index:			<0.001
0–1	8.77%	13.45%	
2–3	30.08%	29.37%	
>3	61.15%	57.17%	
Comorbidities:			
Alcohol use disorder	29.37%	14.98%	<0.001
Bronchiectasis	0.18%	2.26%	<0.001
Cerebrovascular disease	4.66%	10.14%	<0.001
Coronary artery disease	20.53%	16.61%	<0.001
Cystic fibrosis	0.07%	1.05%	<0.001
Drug use disorder	8.79%	7.04%	0.0072
Dyslipidemia	23.56%	15.76%	<0.001
Hypertension	57.78%	45.72%	<0.001
Obesity	15.04%	12.14%	0.0004
Tobacco use disorder	40.86%	23.91%	<0.001
Complications:			
Acute kidney injury	42.96%	73.36%	<0.001
Acute respiratory failure	24.85%	65.74%	<0.001
Bleeding other than GI	1.27%	4.15%	<0.001
Disseminated intravascular coagulation	2.49%	11.46%	<0.001
Gastrointestinal bleeding	20.49%	21.97%	0.1147
Intracranial hemorrhage	1.42%	3.52%	<0.001
Sepsis or infection	52.95%	100.00%	<0.001
Venous thromboembolism [c]	7.99%	16.29%	<0.001
Procedure:			
Invasive mechanical ventilation	18.78%	58.17%	<0.001
Non-invasive ventilation	3.15%	7.30%	<0.001
Length of stay (in days):			<0.001
≤10	80.05%	33.32%	
11–20	13.75%	27.80%	
≥21	6.20%	38.89%	

Table 1. *Cont.*

Variable	LFC Without Aspergillosis (N: 3,005,849)	LFC with Aspergillosis (N: 9515)	*p* Value
Disposition of patient:			<0.001
Routine	50.98%	24.93%	
Transfer to short-term hospital	4.93%	7.26%	
Transfer to a facility [d]	22.94%	45.21%	
Home-health care	18.58%	21.12%	
Left against medical advice	2.57%	1.49%	
Outcomes:			
All-cause inpatient mortality	15.75%	43.40%	<0.001
Length of stay in days, mean (±SD)	7.5 (9.4)	22.9 (24.7)	<0.001
Total hospital charges in US dollars, mean (±SD)	$104,389.4 (200,464.1)	$459,414.9 (682,279.8)	<0.001
Total hospital costs in US dollars, mean (±SD)	$24,272.1 (43,016.1)	$108,030.6 (155,198.1)	<0.001

[a]: Includes Asian or Pacific Islanders, Native Americans, and others. [b]: Includes no charge, worker's compensation, CHAMPUS, CHAMPVA, Title V, and other government programs. [c]: Includes both pulmonary embolism and deep vein thrombosis. [d]: Includes Skilled Nursing Facility (SNF), Intermediate Care Facility (ICF), another type of facility.

3.2. Outcomes

All-cause inpatient mortality was notably higher in the IA group (43.40% vs. 15.75%; $p < 0.001$). Additionally, nearly half of IA admissions required discharge to a nursing or rehabilitation facility (45.21%) compared to 18.58% in the non-aspergillosis group, while routine discharges were significantly more common in the non-aspergillosis group (24.93% vs. 5.98%; $p < 0.001$).

Hospital stays were markedly longer for the IA cohort, with 38.89% of admissions exceeding 21 days compared to 6.20% in the non-aspergillosis cohort ($p < 0.001$). The mean LOS was more than three times longer for IA admissions (22.9 vs. 7.5 days; $p < 0.001$). The mean total hospital charges for the IA cohort were more than four times higher than those for the non-aspergillosis cohort (USD 459,414.9 vs. USD 104,389.4; $p < 0.001$). Similarly, the mean total hospitalization costs in the IA group were over four times greater than in the non-IA group (USD 108,030.6 vs. USD 24,272.1; $p < 0.001$) (Table 1). Table 2 and Figure 1 provide the number of admissions and death of LFC admissions with IA during the study period.

Table 2. Absolute Number of Liver Failure and Cirrhosis Admissions (2016–2020).

	2016	2017	2018	2019	2020	Total
Admissions for liver failure and cirrhosis	544,150	577,480	608,745	637,660	647,330	3,015,364
Admissions for liver failure and cirrhosis with invasive aspergillosis	1960	1740	1820	1810	2185	9515
Death count in admissions for liver failure and cirrhosis with invasive aspergillosis	695	710	825	850	1045	4125

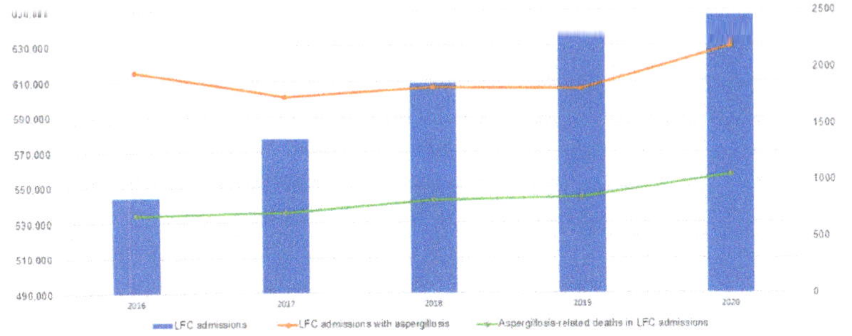

Figure 1. Liver Failure and Cirrhosis Admissions in the US (2016–2020).

4. Discussion

To our knowledge, this is the largest study utilizing a national database to examine the epidemiology and outcomes of IA in patients with liver disease. Fungal infections have long been linked to severe immunosuppression, commonly seen in conditions such as hematologic malignancies, neutropenia, and T-cell suppression [5]. While aspergillosis has traditionally been recognized as an opportunistic infection, recent studies have highlighted its role as a life-threatening complication in patients with liver failure [1,9,10]. A previous study suggests that the increased infection risk in cirrhotic patients may be due to cirrhosis-associated immune dysfunction, which combines immunodeficiency with disturbances in key immune cells, including neutrophils, monocytes, T cells, and B cells [11]. Further supporting this, another study found that both innate and, more notably, adaptive cellular immune impairments emerge early in the progression of liver cirrhosis, potentially increasing susceptibility to infections [12]. A postmortem study revealed that over half of the patients with acute or advanced liver disease had a diagnosis of invasive aspergillosis [1]. The Centers for Disease Control and Prevention data indicate that the incidence of aspergillosis has been rising at an average rate of 3% over the past decade [13]. Given the increasing incidence of aspergillosis and the growing recognition of LFC patients as a high-risk group, our study seeks to address gaps in epidemiological data regarding its prevalence, outcomes, and clinical implications in this vulnerable population.

Between 2016 and 2020, we identified 9515 hospitalizations for LFC and IA, corresponding to an observed incidence of 0.36%. This is similar to findings from a retrospective study in China, which reported an incidence of 0.29% [14]. However, significantly higher rates have been observed in intensive care unit (ICU) settings, reaching 14.28% [10]. Prior research has classified liver cirrhosis patients with an ICU stay of more than 7 days as an intermediate-risk category for invasive aspergillosis, possibly explaining the higher incidence observed in ICUs [2]. Notably, liver transplant recipients have a significantly higher incidence of aspergillosis, exceeding 9.2% [15]. Solid organ transplantation is a well-recognized risk factor for opportunistic aspergillosis, primarily due to prolonged immunosuppressant and corticosteroid exposure. To ensure a more specific analysis, we excluded hospitalizations involving liver transplantation from this study.

The mean age of LFC admissions with IA in our study was 58.4 years, closely aligning with a previous study that reported a mean age of 57 years for LFC patients [16]. We observed a higher incidence of IA among African-American, Hispanic, Asian, Pacific Islander, and other minority populations. Among these groups, Hispanics had the highest proportion of aspergillosis cases (16.27%), underscoring the disproportionately higher burden of aspergillosis in minority communities affected by LFC. This trend is consistent with prior research on COVID-19-associated aspergillosis [17]. Furthermore, LFC admissions with IA were more frequently seen in teaching hospitals compared to those without aspergillosis (83.39% vs. 73.36%). This pattern mirrors findings from a study on influenza-associated aspergillosis and is likely due to the transfer of more critically ill patients—who have a higher risk of developing IA—to academic centers, as well as greater awareness and diagnostic recognition of aspergillosis in teaching hospitals [18].

Our study found that admissions with IA had a younger mean age at admission and a lower Charlson Comorbidity Index (19.18% vs. 29.14% with Charlson Comorbidity Index > 3; $p < 0.001$), suggesting that IA is more closely linked to acute physiological decline rather than cumulative comorbidities. This aligns with the nature of infections, which typically occur as acute events rather than as a consequence of long-term disease burden.

We observed a higher proportion of IA-associated LFC admissions in the wealthiest income quartile, whereas the poorest quartile showed a lower proportion of LFC admissions with IA. This disparity may reflect greater access to healthcare resources and diagnostic tools among higher-income populations, potentially leading to increased detection of IA. However, the database lacks granular information to provide definitive explanation of this observation.

We found that frequency of AKI was significantly higher in the IA cohort compared to the non-IA cohort (73.36% vs. 42.96%). Previous studies have similarly reported that IA contributes to severe organ failure [18,19]. Additionally, we observed a higher incidence of non-gastrointestinal bleeding, disseminated intravascular coagulation, intracranial hemorrhage, and venous thromboembolism in the IA cohort. These complications indicate an increased risk profile that may hinder the feasibility of invasive diagnostic procedures, potentially leading to delays in diagnosis and poorer clinical outcomes. We observed that ARF occurred more than twice as frequently in the IA cohort, a trend consistent with previous studies on IA [18]. This reflects the more critical nature of patients with IA or may be a reason for their increased risk for this infection. Additionally, pulmonary comorbidities such as cystic fibrosis and bronchiectasis were more prevalent in this group. Individuals with underlying lung disease are known to be at increased risk for IA, which typically originates in the lungs and can disseminate to other organs. LFC admissions with IA required IMV over three times more frequently than those without aspergillosis (58.17% vs. 18.78%). Western European studies have reported IMV requirements ranging from 52% to 100% in cirrhotic patients with aspergillosis [10,16]. This high IMV requirement may be due to missed or delayed diagnosis of IA in these non-classical patients, a pattern also observed in previous studies. Diagnostic tools for aspergillosis in non-neutropenic patients, such as serum galactomannan, have limited predictive value, whereas bronchoalveolar lavage specimens have demonstrated greater diagnostic accuracy in this population [9,20]. A recent study highlighted the potential of a non-invasive antibody-guided positron emission tomography and magnetic resonance probe for detecting *Aspergillus fumigatus* lung infections [21]. Given these challenges, diagnosing aspergillosis in non-neutropenic patients requires a comprehensive approach that integrates radiological findings, cultures, fungal biomarkers, and molecular tools for greater accuracy, as also recommended in the structured diagnostic and therapeutic framework proposed by Bassetti et al. [9].

The length of stay for admissions with IA was more than three times longer than that of non-aspergillosis patients (22.9 vs. 7.5 days) and significantly exceeded durations reported in previous large-scale studies, which documented ICU stays of 14 or more days for aspergillosis patients [2,10,22]. The extended hospitalization can likewise be attributed to the atypical presentation and delayed diagnosis of IA in LFC patients.

We found that the mean total hospital charges for the IA cohort were more than four times higher than those for the non-IA cohort (USD 459,414.9 vs. USD 104,389.4). Similarly, mean total hospital[illegible] four times greater in the aspergillosis group compared to the non-aspergillosis group (USD 108,030.6 vs. USD 24,272.1), placing a significant financial strain on the healthcare system. An interesting observation from a study in China highlighted that the high mortality associated with invasive pulmonary aspergillosis was not only due to diagnostic and treatment challenges but also to the heavy economic burden. Many patients, unable to afford the cost of care, ultimately discontinued treatment [3]. These findings emphasize the critical need for early diagnosis, improved treatment strategies, and cost-effective interventions to reduce both clinical and financial burdens.

Mortality rates for aspergillosis in cirrhotic patients have remained alarmingly high, with prior studies reporting rates exceeding 50% [1–3]. A cohort study from China observed mortality surpassing 70%, while a decade-long analysis from France documented a lower hospital mortality rate of 37% [3,16]. Additionally, a large-scale study on ICU patients with cirrhosis and aspergillosis has reported a 100% mortality rate [10]. In our study, we found a mortality rate of 43%, aligning more closely with the French data. Compared to neutropenic patients, non-neutropenic individuals are significantly less likely to exhibit classic symptoms of IA, making early detection more challenging [23]. Furthermore, a large study examining both neutropenic and non-neutropenic patients found that classic radiographic signs, such as nodules and cavitation, were rare. Instead, non-specific findings like consolidation, ground-glass infiltrates, and pleural effusions were more commonly observed [23]. This lack of distinct imaging features further complicates the diagnosis of aspergillosis in LFC patients. The combination of delayed diagnosis and atypical presentation often leads to aspergillosis being recognized only at critical stages, contributing to poorer outcomes. Another important factor that may play a role in the outcome of aspergillosis in this patient population is the challenge with starting and maintaining antifungal therapy given the increased risk of liver dysfunction associated with these medications. The NIS database does not provide information about antifungal therapy in these patients.

To our knowledge, this is the largest study examining aspergillosis as a complication in liver failure patients. It provides valuable insights into its incidence and outcomes; however, several limitations must be acknowledged.

One of the limitations is the lack of data on diagnostic tools and treatment strategies, as the NIS database does not capture this information. Specifically, the database does not provide information on the etiology of LFC, treatment such corticosteroids or other immunomodulatory medications, or antifungal therapy. Additionally, the structure of the NIS prevents tracking individual patients across multiple admissions, with each entry representing a separate hospitalization rather than a continuous patient history. This restriction limits longitudinal analysis and may obscure disease progression and treatment pathways [24]. Furthermore, the NIS lacks the clinical information needed to apply the European Organization for Research and Treatment of Cancer criteria for IA, including host factors indicative of immunocompromised status. Given the low observed incidence and the inability to exclude alternative diagnoses such as chronic pulmonary aspergillosis or to identify established host risk factors, establishing a definitive association or causal relationship between LFC and IA is challenging. Moreover, since the timing of diagnoses cannot be determined in the NIS, it remains unclear whether IA represents a pre-existing condition, the cause of admission, or a complication arising during hospitalization.

While national databases offer valuable epidemiological insights, findings should be interpreted cautiously due to the risk of misclassification bias and inconsistencies from coding errors and variations in documentation practices across healthcare facilities. Given these limitations, generalizing results from database studies requires careful consideration.

This study, the largest to date, highlights the significant burden of IA in LFC patients who experience worse clinical outcomes, prolonged hospital stays, and higher healthcare costs. Diagnosing invasive aspergillosis remains challenging due to lack of awareness, the atypical presentation, and the limited sensitivity and specificity of diagnostic studies. A comprehensive diagnostic approach integrating imaging, cultures, fungal biomarkers, and molecular tools is crucial for early detection and timely intervention.

These findings underscore the urgent need for targeted interventions, interdisciplinary collaboration, and cost-effective strategies to improve outcomes in this high-risk yet underrecognized population. Strengthening diagnostic accuracy and optimizing treatment approaches are essential to reducing both clinical and economic burdens.

Supplementary Materials: The following supporting information can be downloaded at: https://www.mdpi.com/article/10.3390/jof11050334/s1, Table S1: The quartile classification of the estimated median household income of residents in the patient's ZIP code vary by every year. Listed below are the dollar ranges for study duration 2017–2019.

Author Contributions: A.S.: In planning, conducting the study, collecting and interpreting data, and drafting the manuscript. B.M.: In interpreting data and drafting the manuscript. A.O.S.: In planning, conducting the study, collecting and interpreting data, and drafting the manuscript. All authors have read and agreed to the published version of the manuscript.

Funding: This research received no external funding.

Institutional Review Board Statement: Not applicable.

Informed Consent Statement: Not applicable.

Data Availability Statement: The original contributions presented in the study are included in the article/Supplementary Materials, further inquiries can be directed to the corresponding author.

Conflicts of Interest: The authors declare no conflicts of interest.

Abbreviations

AKI	Acute kidney injury
ARF	Acute respiratory failure
IA	Invasive aspergillosis
ICD-10-CM	International Classification of Disease, Tenth Revision, Clinical Modification
ICD-10-PCS	International Classification of Diseases, Tenth Revision, Procedure Coding System
ICU	Intensive care unit
IMV	Invasive mechanical ventilation
LFC	Liver failure and cirrhosis
LOS	Length of stay
NIS	National inpatient sample
NIV	Non-invasive ventilation
SD	Standard deviation

References

1. Falcone, M.; Massetti, A.P.; Russo, A.; Vullo, V.; Venditti, M. Invasive aspergillosis in patients with liver disease. *Med. Mycol.* **2011**, *49*, 406–413. [CrossRef] [PubMed]
2. Meersseman, W.; Lagrou, K.; Maertens, J.; Van Wijngaerden, E. Invasive aspergillosis in the intensive care unit. *Clin. Infect. Dis. Off. Publ. Infect. Dis. Soc. Am.* **2007**, *45*, 205–216. [CrossRef]
3. Zhang, X.; Yang, M.; Hu, J.; Zhao, H.; Li, L. Epidemiology of invasive pulmonary aspergillosis in patients with liver failure: Clinical presentation, risk factors, and outcomes. *J. Int. Med. Res.* **2018**, *46*, 819–827. [CrossRef] [PubMed]
4. Vallabhaneni, S.; Benedict, K.; Derado, G.; Mody, R.K. Trends in Hospitalizations Related to Invasive Aspergillosis and Mucormycosis in the United States, 2000–2013. *Open Forum Infect. Dis.* **2017**, *4*, ofw268. [CrossRef]

5. Kousha, M.; Tadi, R.; Soubani, A.O. Pulmonary aspergillosis: A clinical review. *Eur. Respir. Rev.* **2011**, *20*, 156–174. [CrossRef]
6. Healthcare Cost and Utilization Project—National (Nationwide) Inpatient Sample (HCUP-NIS)—Healthy People 2030 [illegible]. Available online: https://health.gov/healthypeople/objectives-and-data/data-sources-and-methods/data-sources/healthcare-cost-and-utilization-project-national-nationwide-inpatient-sample-hcup-nis (accessed on 11 January 2024).
7. Khalifa, A.; Obeid, J.S.; Gregoski, M.J.; Rockey, D.C. Accurate Identification of Patients with Cirrhosis and Its Complications in the Electronic Health Record. *Dig. Dis. Sci.* **2023**, *68*, 2360–2369. [CrossRef] [PubMed]
8. Overview of the National (Nationwide) Inpatient Sample (NIS). Available online: https://hcup-us.ahrq.gov/nisoverview.jsp (accessed on 19 June 2023).
9. Bassetti, M.; Peghin, M.; Vena, A. Challenges and Solution of Invasive Aspergillosis in Non-neutropenic Patients: A Review. *Infect. Dis. Ther.* **2018**, *7*, 17–27. [CrossRef]
10. Lahmer, T.; Brandl, A.; Rasch, S.; Baires, G.B.; Schmid, R.M.; Huber, W.; Mayr, U. Prevalence and outcome of invasive pulmonary aspergillosis in critically ill patients with liver cirrhosis: An observational study. *Sci. Rep.* **2019**, *9*, 11919. [CrossRef]
11. Albillos, A.; Lario, M.; Álvarez-Mon, M. Cirrhosis-associated immune dysfunction: Distinctive features and clinical relevance. *J. Hepatol.* **2014**, *61*, 1385–1396. [CrossRef] [PubMed]
12. Rueschenbaum, S.; Ciesek, S.; Queck, A.; Widera, M.; Schwarzkopf, K.; Brüne, B.; Welsch, C.; Wedemeyer, H.; Zeuzem, S.; Weigert, A.; et al. Dysregulated Adaptive Immunity Is an Early Event in Liver Cirrhosis Preceding Acute-on-Chronic Liver Failure. *Front. Immunol.* **2021**, *11*, 534731. [CrossRef]
13. CDC Data and Statistics on Aspergillosis. Aspergillosis. 23 April 2024. Available online: https://www.cdc.gov/aspergillosis/statistics/index.html (accessed on 19 January 2025).
14. Chen, J.; Yang, Q.; Huang, J.; Li, L. Clinical findings in 19 cases of invasive pulmonary aspergillosis with liver cirrhosis. *Multidiscip. Respir. Med.* **2014**, *9*, 1. [CrossRef] [PubMed]
15. Husain, S.; Camargo, J.F. Invasive Aspergillosis in solid-organ transplant recipients: Guidelines from the American Society of Transplantation Infectious Diseases Community of Practice. *Clin. Transplant.* **2019**, *33*, e13544. [CrossRef] [PubMed]
16. Levesque, E.; Ait-Ammar, N.; Dudau, D.; Clavieras, N.; Feray, C.; Foulet, F.; Botterel, F. Invasive pulmonary aspergillosis in cirrhotic patients: Analysis of a 10-year clinical experience. *Ann. Intensive Care* **2019**, *9*, 31. [CrossRef] [PubMed]
17. Sharma, A.; Sharma, A.; Soubani, A.O. Incidence and inhospital outcomes of coronavirus disease 2019-associated pulmonary aspergillosis in the United States. *Ann. Thorac. Med.* **2024**, *19*, 87–95. [CrossRef] [PubMed]
18. Sharma, A.; Mishra, T.; Kumar, N.; Soubani, A.O. Influenza-Associated Aspergillosis: Nationwide Trends, Predictors and Outcomes from 2005 to 2014. *Chest* **2020**, *158*, 1857–1866. [CrossRef] [PubMed]
19. Taccone, F.S.; Van den Abeele, A.M.; Bulpa, P.; Misset, B.; Meersseman, W.; Cardoso, T.; Paiva, J.A.; Blasco-Navalpotro, M.; De Laere, E.; Dimopoulos, G.; et al. Epidemiology of invasive aspergillosis in critically ill patients: Clinical presentation, underlying conditions, and outcomes. *Crit. Care* **2015**, *19*, 7. [CrossRef] [PubMed]
20. Zhou, W.; Li, H.; Zhang, Y.; Huang, M.; He, Q.; Li, P.; Zhang, F.; Shi, Y.; Su, X. Diagnostic Value of Galactomannan Antigen Test in Serum and Bronchoalveolar Lavage Fluid Samples from Patients with Nonneutropenic Invasive Pulmonary Aspergillosis. *J. Clin. Microbiol.* **2017**, *55*, 2153–2161. [CrossRef]
21. Rolle, A.M.; Hasenberg, M.; Thornton, C.R.; Solouk-Saran, D.; Männ, L.; Weski, J.; Maurer, A.; Fischer, E.; Spycher, P.R.; Schibli, R.; et al. ImmunoPET/MR imaging allows specific detection of Aspergillus fumigatus lung infection in vivo. *Proc. Natl. Acad. Sci. USA* **2016**, *113*, E1026–E1033. [CrossRef]
22. Garbino, J.; Fluckiger, U.; Elzi, L.; Imhof, A.; Bille, J.; Zimmerli, S. Survey of aspergillosis in non-neutropenic patients in Swiss teaching hospitals. *Clin. Microbiol. Infect. Off. Publ. Eur. Soc. Clin. Microbiol. Infect. Dis.* **2011**, *17*, 1366–1371. [CrossRef]
23. Cornillet, A.; Camus, C.; Nimubona, S.; Gandemer, V.; Tattevin, P.; Belleguic, C.; Chevrier, S.; Meunier, C.; Lebert, C.; Aupée, M.; et al. Comparison of epidemiological, clinical, and biological features of invasive aspergillosis in neutropenic and nonneutropenic patients: A 6-year survey. *Clin. Infect. Dis. Off. Publ. Infect. Dis. Soc. Am.* **2006**, *43*, 577–584. [CrossRef]
24. Khera, R.; Krumholz, H.M. With Great Power Comes Great Responsibility: "Big Data" Research from the National Inpatient Sample. *Circ. Cardiovasc. Qual. Outcomes* **2017**, *10*, e003846. [CrossRef] [PubMed]

Journal of *Fungi*

MDPI

Review

Management of Intra-Abdominal Candidiasis in Intensive Care Setting: A Narrative Review

Marco Marotta Pais [1,2], Rafael Zaragoza [3], Ignacio Martín-Loeches [4], Frederic F. Gómez-Bertomeu [2,5,6,7] and Alejandro Rodríguez [1,6,8,9,*]

1 Critical Care Department, Hospital Universitari de Tarragona Joan XXIII, Mallafré Guasch 4, 43007 Tarragona, Spain; mmarottapais@gmail.com
2 Department of Medicine and Surgery, Faculty of Medicine and Health Sciences, Rovira & Virgili University, 43005 Tarragona, Spain; ffgomez.hj23.ics@gencat.cat
3 Critical Care Department, Hospital Universitario Dr. Peset, Av. Gaspar Aguilar 90, 46017 Valencia, Spain; zaragoza_raf@gva.es
4 Department of Intensive Care Medicine, Multidisciplinary Intensive Care Research Organization (MICRO), St James' Hospital, D08 NHY1 Dublin, Ireland; drmartinloeches@gmail.com
5 Microbiology/Clinical Analysis Laboratory, Hospital Universitari de Tarragona Joan XXIII, Mallafré Guasch 4, 43007 Tarragona, Spain
6 IISPV (Instituto de Investigación Sanitaria Pere Virgili), 43005 Tarragona, Spain
7 Centre for Biomedical Research in Infectious Diseases Network (CIBERINFEC), 28220 Madrid, Spain
8 Centre for Biomedical Research Network Respiratory Diseases (CIBERES), 43005 Tarragona, Spain
9 Department of Basic Medical Sciences, Faculty of Medicine and Health Sciences, Rovira & Virgili University, 43201 Reus, Spain
* Correspondence: ahr1161@yahoo.es

Abstract: Intra-abdominal candidiasis (IAC), with or without candidemia, is a common condition in patients in intensive care units (ICUs). Early diagnosis of IAC remains a challenge for clinicians despite new biomarkers. Early and appropriate antifungal treatment, which is associated with better clinical outcomes, is negatively affected by the increased isolation of non-albicans *Candida* strains that are resistant to the commonly used azoles and echinocandins. Based on the pharmacokinetic (PK) and pharmacodynamic (PD) properties of the different treatment options, liposomal amphotericin B, rezafungin or high doses of anidulafungin appear to be the most appropriate first-line options for complicated IAC in ICUs.

Keywords: invasive candidiasis; antifungal treatment; diagnosis; outcome

1. Introduction

Intra-abdominal candidiasis (IAC), with or without associated candidemia, is caused by overgrowth of *Candida* spp. in the abdominal cavity (see Table 1), and its incidence has increased in recent decades. IAC mainly affects critically ill patients undergoing major abdominal surgery, especially those with complex surgical cases or immunocompromised individuals [1–5].

Despite advances in prevention strategies, diagnostic tools and therapeutic approaches, IAC remains associated with high mortality rates, reaching up to 60% [6,7] This highlights the critical need for ongoing evaluation and optimization of management strategies [1–5].

The aim of this review is to update key concepts related to the changing epidemiology of IAC towards non-albicans *Candida* strains and the impact this situation may have on appropriate antifungal treatment, as well as to update concepts on the impact of *Candida* spp.

in the peritoneal cavity, the risk factors associated with its presence and the interactions with bacteria leading to increased mortality. Finally, basic concepts of [illegible] antifungal treatment strategies based on pharmacokinetic/pharmacodynamic (PK/PD) parameters and antifungal susceptibility will be developed in order to develop an empirical treatment algorithm adapted to clinical realities.

Table 1. Classification of intra-abdominal candidiasis (IAC) types (modified from reference [8]).

Primary Peritonitis	Peritoneal inflammation associated with the isolation of *Candida* spp. in the absence of gastrointestinal perforation or any other visceral process that justifies it.
Secondary Peritonitis	Peritoneal infection by *Candida* spp. because of a pathological process or gastrointestinal perforation.
Intra-Abdominal Abscess of Gastrointestinal Origin	Isolation of *Candida* spp. in a collection of purulent material arising from secondary peritonitis.
Secondary Peritonitis of Hepatobiliary or Pancreatic Origin	Peritoneal infection by *Candida* spp. resulting from a pathological process of the liver, gallbladder, bile ducts or pancreas.
Intra-Abdominal Abscess of Hepatobiliary or Pancreatic Origin	Abscess with isolation of *Candida* spp. resulting from a pathological process of the liver, gallbladder, bile ducts or pancreas. Infected bilomas, pancreatic pseudocysts or other (peri)pancreatic collections are also classified as abscesses.
Infected Pancreatic Necrosis	Isolation of *Candida* spp. in non-viable pancreatic tissue as a consequence of pancreatitis.
Cholecystitis, Cholangitis	Isolation of *Candida* spp. in material from the gallbladder or biliary tract.
Recurrent IAC	Isolation of *Candida* spp. appearing after the apparent resolution of clinical or radiological findings of an initial infection.
Persistent IAC	Isolation of *Candida* spp. in a new intra-abdominal sample 48 h after adequate source control and appropriate antifungal treatment.

2. Epidemiology

There are at least 15 different *Candida* species that cause human disease, but over 95% of invasive disease is caused by the 6 most common pathogens: *Candida albicans*, *Candida glabrata*, *Candida tropicalis*, *Candida parapsilosis*, *Candida krusei* and, in some regions, *Candida auris*. Severe infections due to these organisms are collectively referred to as invasive candidiasis (IC) [8]. In recent decades, there has been a notable epidemiological shift with an increase in non-albicans *Candida* species. This shift from *C. albicans* to non-albicans species has significant clinical implications, particularly in terms of treatment efficacy, patient survival and antifungal resistance. Non-albicans species such as *C. glabrata*, *C. krusei*, *C. tropicalis* and *C. auris* often exhibit intrinsic or acquired resistance to typical antifungal agents. *C. glabrata* and *C. krusei* often show reduced susceptibility to azoles and require the use of echinocandins or higher doses of alternative agents, which may not always be feasible in critically ill or renally compromised patients [9,10].

Differences in virulence, tissue tropism and biofilm-forming ability among non-albicans species complicate both diagnosis and eradication, leading to persistent or recurrent infections despite antifungal treatment [9–11]. Of particular note is a relatively new species, *Candida auris*, which has become a global health concern. Since it was first identified in Japan (2009), it has spread rapidly to more than 35 countries [12–14]. The World Health Organization (WHO) has classified *C. auris* as a critical priority pathogen on its fungal priority list. The significance of *C. auris* lies in its high resistance to antifungal agents, which severely limits therapeutic options. In addition, the challenges associated with its laboratory identification, combined with its ease of transmission, provide ideal conditions for epidemic outbreaks. Furthermore, *C. auris* can persist in the environment for weeks, and the lack of effective decolonization methods makes controlling its spread a major challenge [13,15]. This shift is also correlated with variable and increased mortality rates of up to 40–60% [15]. Studies have shown that infections caused by non-candida species, especially *C. glabrata* and *C. auris*, are associated with delays in appropriate antifungal therapy, which is a critical factor, and higher 30-day mortality compared to *C. albicans* [16,17].

The distribution of *Candida* species varies significantly between regions, with direct implications for treatment strategies. In North America and Northern Europe, *C. glabrata* is more prevalent and, as we mentioned before, has reduced susceptibility to fluconazole, making echinocandins the preferred first-line therapy. In Spain, non-*albicans Candida* species are increasingly implicated in candidemia. A multicenter study (CANDIMAD) [18] from 2020 to 2022 reported a rise in *C. parapsilosis* from 21.6% to 36.1%, alongside a decrease in *C. albicans*. Notably, fluconazole-resistant *C. parapsilosis* strains, particularly the CP-451 genotype, have become endemic in several hospitals.

Latin America and parts of Asia report a higher prevalence of *C. parapsilosis*, which tends to be more susceptible to azoles but has higher minimum inhibitory concentrations (MICs) for echinocandins. Meanwhile, *C. tropicalis* is particularly common in Southeast Asia and India, often associated with hematological malignancies and high virulence [19].

Regarding IAC, Dupont H et al. analyzed samples from patients with peritonitis admitted to the ICU, of whom 30.6% had *Candida* spp. Peritonitis [20]. The most frequently isolated species in these cases was *C. albicans* (74%), followed by *C. glabrata* (17%). Vergidis P et al. [21] identified 163 IAC cases and 161 candidemias among more than 34,000 patients, with a rate of 4.7 episodes per 1000 admissions. A multicenter study (EUCANDICU) conducted in 23 ICUs across nine European countries analyzed 570 episodes of invasive candidiasis (7 episodes per 1000 ICU admissions). Of these cases, 65% were candidemia, 29% were IAC and 5% were IAC in combination with candidemia. A noteworthy finding of this study is that 43% of cases involved non-albicans *Candida* species, with an 8% rate of co-infection by two species [2].

As is evident, the epidemiology of both candidemia and IAC has shifted towards non-albicans *Candida* species. The reasons for this shift are thought to be linked to the increasing use of azoles for "prophylactic" purposes, as well as the rise in gastrointestinal pathology among older patients with more comorbidities. These findings should be carefully considered when determining the empirical treatment for these patients. The overall quality of the available evidence remains poor and uniform recommendations cannot be made.

3. Risk Factors

As outlined in various publications by different authors, there are different risk factors associated with the likelihood of invasive candidiasis [1–5,8]. There are those patient-related, treatment-related and procedural-related factors. Table 2 presents the most common risk factors. Prior colonization, especially multifocal colonization, is a crucial factor that should be assessed in critically ill patients, as it plays a key role in diagnosing or predicting IAC using various scoring systems. Multifocal colonization should be actively sought in all critical patients with an ICU stay of more than 7–10 days, especially if they have risk factors.

Table 2. Risk factors for invasive candidiasis in non-neutropenic adult patients (references: [1–5,8]).

Adults		
Patient-Dependent	**Treatment-Related**	**Procedure-Related**
Chronic renal failure	Broad-spectrum antibiotics	Abdominal surgery
Severe pancreatitis	Antifungal prophylaxis	Mechanical ventilation
High disease severity	Corticosteroids	Central venous catheter
Diabetes	Parenteral nutrition	
Multifocal colonization	Hemodialysis	
Intra-abdominal bacterial infection	Prolonged ICU stay	

Another important factor is prior antibiotic treatment, which is considered a significant risk factor for IAC in nearly all studies. It is important to note that approximately 60% of ICU patients receive antibiotic treatment for various reasons [22], which significantly increases the number of patients at potential risk for IAC. Restrictive policies on the use of antimicrobials and a reduction in the duration of treatments are measures that could be taken to reduce the risk of IAC.

Finally, the widespread overuse of fluconazole for preventive or prophylactic purposes exerts selective pressure not only on *Candida* spp. but also markedly contributes to their resistance to azoles.

Research has identified comparable risk factors for IAC. Bassetti et al. published a study on predictive factors for IAC in ICU patients, using a logistic regression analysis from the EUCANDICU study [23]. The authors observed that recurrent gastrointestinal perforation (OR = 13.9, 95% CI 2.6–72.8), anastomotic leakage (OR = 6.6, 95% CI 1.9–21.9), intra-abdominal drains (OR = 6.5, 95% CI 1.7–25.0), receiving antifungal treatment (OR = 4.2, 95% CI 1.04–17.4) or antibiotic treatment for more than seven days (OR = 3.7, 95% CI 1.3–10.5) were independent predictors. As Bassetti et al. highlighted in another multicenter study involving more than 300 cases of candidemia, intra-abdominal infection increases the risk of developing shock (OR = 2.18, 95% CI 1.04–4.55) [24]. Dupont H et al. used a multivariate model to assess whether the presence of *Candida* spp. could be predicted in critically ill patients with peritonitis. The authors found that a high gastrointestinal origin (above the Treitz ligament) increased the risk of *Candida* isolation by 2.4 times (OR = 2.38), as did antibiotic administration (OR = 2.25) and the presence of shock (OR = 2.45) [20].

As antibiotic treatment is mandatory in patients with peritonitis, it remains a non-modifiable risk factor. However, it should be specifically considered when deciding whether to initiate antifungal therapy. Furthermore, as will be discussed later, in more than 40% of cases, bacterial and fungal infections coexist, and this association appears to have a significant impact on patient mortality [25,26].

Interactions between *C. albicans* and gut bacteria can be either antagonistic or synergistic. Species like *Pseudomonas aeruginosa*, *Salmonella* spp. and *Lactobacillus* exert inhibitory effects on *C. albicans*. Understanding these dynamics is crucial for developing targeted probiotic or microbiota-modulating therapies to prevent overgrowth and invasive intestinal candidiasis [27]. There are still many aspects of the pathogenesis of IAC that are not fully understood. One area of increasing interest is the role of gut enterotypes and patient response. These enterotypes are largely influenced by the host's diet and shape the metabolic landscape of the gut microbiota. Enterotype I bacteria derive their energy mainly from carbohydrates through glycolysis and the pentose phosphate pathway. In contrast, enterotypes II and III harbor microbial populations capable of degrading glycoproteins present on the mucosal surface. This activity can weaken the epithelial barrier, promote immune evasion and create a microenvironment conducive to fungal colonization and overgrowth. Specifically, this alteration may predispose to the invasion of *Candida* spp. and contribute to the initiation or progression and lack of resolution of intra-abdominal infections [28]

4. Diagnosis

The isolation of *Candida* spp. in blood (candidemia) or other sterile body fluids is a clear criterion for infection. However, the diagnostic significance of its isolation in respiratory samples (which has no significance except in immunocompromised patients) or in samples obtained by puncture/surgery (especially those obtained from abdominal drains, which are potentially contaminated) remains a subject of controversy.

The initial diagnostic approach involves the identification of patients with the previously mentioned risk factors. Secondly, it is essential to obtain intra- or perioperative samples (fluid or tissue), as well as any other clinically relevant samples. Surveillance cultures also play an important role in diagnosis by demonstrating multiple colonization (Table 3). Additionally, predictive scores (see Table 4) can be applied to patients with more than two weeks of ICU stay, allowing us to assess the risk probability of invasive candidiasis.

Thirdly, the serological determinations of different biomarkers are also considered, although these are not available in many centers and their diagnostic capacity is still under discussion.

The gold standard for diagnosing invasive candidiasis is blood culture, although this method has significant limitations. Its sensitivity ranges from 50% to 75%, and the time required from sample collection to fungal identification is at least 48–72 h. This is particularly problematic because yeast replication is often slower than bacterial replication, leading to delayed blood culture positivity. Once a blood culture has returned a positive result, further identification is required [29,30].

Table 3. Different diagnostic techniques for invasive candidiasis.

Methods	Comments	Performance	Key Points
Detection of mannan antigen (M-Ag) and anti-mannan antibody (M-Ac)	Combined detection of antigen and antibody is recommended for all patients suspected of invasive candidiasis.	Se: 80–90% Sp: 85–90% PNV: 85–95% PPV: 70–85% [31]	Sensitivity varies depending on *Candida* species (e.g., lower for *C. krusei* or *C. parapsilosis*); limited utility in neutropenic patients or those with strong immunosuppression.
Detection and quantification of anti-mycelial or anti-germ tube antibodies (CAGTAs)	When antibody titers exceed 1:160. Additionally, it may help distinguish between transient or catheter-related candidemia (CAGTA−) and deep-seated infection-associated candidemia (CAGTA+).	Se: 85% Sp: 95% PPV: 72–89% NPV: 92–97% [32]	A negative CAGTA result has a high NPV, meaning it is especially useful to rule out invasive candidiasis in high-risk patients.
Determination of 1,3-ß-D-glucan	1,3-ß-D-glucan is a fungal cell wall component released during infection. It is a non-specific biomarker and does not identify the causative species.	Se: 70–90% Sp: 75–95% PPV: 50–80% (depends on pre-test probability). NPV: 95–100% in high-risk populations [33]	A negative BDG test is particularly valuable for ruling out invasive candidiasis (high NPV), especially in ICU and immunocompromised patients. It is valuable for discontinuing antifungal treatment.

Table 3. Cont.

Methods	Comments	Performance	Key Points
In situ hybridization PNA-FISH Yeast Traffic Light	Uses species-specific peptide nucleic acid (PNA) probes that hybridize in situ with the target DNA of various *Candida* species, producing fluorescence in different colors. It also allows the identification of fluconazole-sensitive species, facilitating treatment adjustments and potential de-escalation.	Se: 92–100% Sp: 95–100%. PPV: >98% NPV: >98% [34]	Especially valuable where *C. glabrata* (often fluconazole-resistant) must be differentiated from *C. albicans* early. Rapid technique.
Mass spectrometry with soft ionization (MALDI-TOF)	This technique is performed on a positive blood culture. It identifies bacteria and fungi through proteomic analysis, allowing accurate identification of fungal isolates in solid culture media in less than 30 min.	Se: 95–100% Sp: 98% PPV: 95% NPV 95% [35]	Highly useful for early targeted therapy when paired with automated blood culture systems.
Detection of nucleic acids via polymerase chain reaction (PCR)	This technique enables the detection of bacterial DNA and certain *Candida* species in blood or sterile fluids using real-time multiplex PCR or microarrays. The limitation is that it requires a positive blood culture, but it allows rapid detection (1 h).	Se: 80–95% Sp: 90–98% PPV: up to 90% NPV: 95% [33]	Early and direct detection of *Candida* DNA from blood or sterile fluids, often before cultures become positive. Particularly useful in high-risk populations (e.g., ICU, neutropenia and post-surgery). False positives may occur due to detection of colonization or non-viable DNA.
T2 Rm (Candida panel)	This technique detects *Candida* spp. using magnetic resonance within 3–5 h.	Se: 91–95% Sp: 98–99% PPV: 70–80%% NPV: >99% [36]	Allows species-level diagnosis within hours—much faster than cultures (48–72 h). Useful in patients with suspected candidemia, but negative cultures can reduce empirical antifungal use and improve targeted therapy. Unfortunately, this device has been withdrawn from the market in Europe.

Sensitivity (Se); Specificity (Sp); Positive predictive value (PPV); Negative predictive value (NPV).

4.1. Biomarkers for Diagnosing Invasive Candidiasis

León C et al. evaluated the usefulness of different biomarkers in critically ill patients with intra-abdominal pathology, both individually and in combination. The study's key findings included the positive results for 1,3-ß-D-glucan (BDG) and anti-mycelial or anti-germ tube antibodies (CAGTAs) in a single sample, or at least one of them in two consecutive samples, which was identified as the most effective strategy for diagnosing invasive candidiasis (IC). In contrast, the individual or combined positivity of C-PCR, mannan antigen (M-Ag) and anti-mannan antibody (M-Ac) was found to have minimal value in distinguishing IC from colonization. The authors concluded that BDG is the most suitable biomarker for diagnosing IC in critically ill patients with acute abdominal complications. Its diagnostic performance is enhanced when combined with CAGTAs [37]. However, it should be noted that other studies have not consistently replicated these findings.

Macrophage-derived CD36+ exosomes have emerged as promising non-invasive biomarkers for Candida albicans infection, showing specificity and elevated levels during fungal exposure. Their unique immunomodulatory content and detectability in biological fluids suggest diagnostic potential for early infection detection and monitoring, surpassing the limitations of conventional diagnostic methods [38].

In our opinion, and in agreement with the majority of researchers, BDG is the biomarker that should be requested in clinical practice, when available, in patients at high risk of invasive candidiasis. Although antifungal treatment should be initiated empirically and early in critically ill and unstable patients with risk factors for candidemia, without waiting for biomarker results, the determination of BDG may contribute to the diagnosis and possibly to the continuation or withdrawal of treatment, given its high negative predictive value.

4.2. Predictive Scores for Invasive Candidiasis

Several researchers have attempted to develop the most effective predictive scores to assess the risk of invasive candidiasis in patients with risk factors. Table 4 provides a comprehensive overview of the main predictive scores, their characteristics and limitations. One of the earliest predictive scales was the Pittet colonization index [39], which was designed for surgical patients. While it is a useful tool, it requires quantitative cultures, which can put a strain on microbiology services. This study demonstrated that a corrected colonization index greater than 0.4 had up to a 100% positive and negative predictive value. León C et al. conducted a prospective study involving 1720 critically ill patients with an ICU stay of over 7 days, leading to the development of the "*Candida* score" or "Sevilla Score". A score greater than 2.5 was able to predict *Candida* infection with 81% sensitivity and 74% specificity [40]. This score was validated in a subsequent prospective study of 1107 critically ill, non-neutropenic patients, confirming that a *Candida* score ≥3 was useful for distinguishing infection. Its diagnostic accuracy was superior to the Pittet colonization index. It is important to highlight that prior colonization by *Candida* spp. is a mandatory criterion for a *Candida* score > 3 [41].

Table 4. Predictive scales for invasive candidiasis used in clinical practice.

Scale	Variables	Cutoff Point	Sensitivity/Specificity	Comments
Pittet Colonization Index [39]	Ratio of the number of colonized sites to the number of cultured sites	≥0.5	100%/69%	In surgical patients, it requires quantitative cultures from all sites, which increases cost and microbiology processing time.
Corrected Pittet Colonization Index [39]	Ratio of the number of sites with high colonization (>10^5 CFU) to the number of cultured sites	≥0.4	100%/100%	Similar to the previous index.
Candida Score [40]	Multifocal colonization (1 point) Abdominal surgery (1 point) Parenteral nutrition (1 point) Severe sepsis—shock (2 points)	≥3	77%/66%	More practical for ICU patients with >7 days of stay. Note that with the cutoff point of 3, multifocal colonization is mandatory.
Nebraska Rule [42]	Stay > 4 days in ICU + • Broad-spectrum antibiotics • Presence of CVC • Abdominal surgery • Steroid treatment • Parenteral nutrition • Average stay before ICU	2.45	84%/60%	More complex to implement as it requires performing a regression calculation with coefficients of the different risk factors.

As discussed later, different authors have implemented these scales to enable early treatment, observing a decrease in the incidence of invasive candidiasis, though often not significant, with no impact on mortality. In our opinion, patients with a prolonged stay in the ICU, risk factors and a *Candida* score > 3 should be considered highly suspicious for invasive candidiasis (provided there is multifocal colonization) and should be started on antifungal treatment early. However, other circumstances that may increase the risk of fungal infection should be considered. In particular, patients with decompensated cirrhosis, hepatitis and those who have undergone liver transplantation are at high risk of developing *Candida* infections. Many factors, such as host immune dysfunction, blood–brain barrier dysfunction, malnutrition and microbiome alterations, contribute to this increased risk of developing fungal infections.

In these patients with a high pretest probability of invasive candidiasis, and where available, it would be appropriate to use molecular diagnostic techniques or biomarker determination (BDG) to refine the diagnosis and avoid overuse of antifungals.

5. Mortality

The mortality rate for invasive candidiasis is high, ranging from 40% to 60% according to various authors [1–5,8,43]. The mortality rate for transient candidemia (e.g., secondary to catheter use) is significantly lower than that for candidemia caused by a deep focus (e.g., endocarditis and intra-abdominal infections). Researchers have explored the impact of early antifungal treatment (see below) and have not observed a favorable effect on mortality in patients with invasive candidiasis. It is crucial to consider three key factors in the pathogenesis of this condition to understand its high mortality.

Firstly, a high fungal load may be a result of broad-spectrum antibiotic use, especially anti-anaerobic antimicrobials. This elevated load or colonization facilitates the translocation of the yeast from the gastrointestinal tract in patients who commonly suffer from prolonged ileus and severe intestinal microcirculation alterations.

Secondly, the disruption of the integrity of both the cutaneous–mucosal barrier (candidemia) and the gastrointestinal tract (IAC) must be considered, whether due to spontaneous perforations or prolonged use of drains or chronic inflammatory processes (e.g., pancreatitis).

Finally, the third element is the dysfunction of immunity that every critically ill, non-neutropenic patient experiences, leading to dissemination to deep tissues [44]. In this context, it is important to note that tertiary peritonitis has a poor prognosis beyond treatment due to immune dysfunction in the peritoneum, which hinders the resolution of the infectious process.

Despite acknowledging these factors, the positive impact of appropriate antifungal treatment is indisputable. In the initial study by Morrel L et al., early treatment (within 12 h) was associated with a significant reduction in mortality (12%) compared to late treatment (>12 h), which had a 30% mortality rate in patients with candidemia [44]. This finding has been replicated by other researchers [45,46]. In their multivariate analysis, Morrel L. et al. [36] found that treatment delay was associated with twice the risk of death (OR = 2.06). In the same study, prior antibiotic administration was found to be associated with a higher mortality (OR = 4.06). Garey K et al. [40] found that fluconazole treatment in patients with candidemia was associated with lower mortality (15%) when administered on day 1 (mortality 23%), day 2 (mortality 37%) or day 3 (mortality 42%) after blood culture. These authors reported a 1.5 higher risk of mortality (OR = 1.52) for each day of treatment delay [46]. Finally, Zilberberg MD et al. reported a higher mortality rate (28.9%, $p = 0.056$)

in patients with candidemia who received inappropriate treatment. It is noteworthy that in this study mortality was 0% in patients with appropriate treatment [47].

In an observational, prospective, multicenter study in critically ill patients with community-acquired intra-abdominal infection, the isolation of *Candida* spp. in peritoneal samples was closely associated with mortality [48]. Despite the introduction of new antifungals, such as echinocandins in the early 2000s, mortality in IAC remains above 50%. The shift in IAC etiology towards non-albicans species is partly responsible for this high mortality, along with increased empirical inappropriate treatment due to high resistance to commonly used antifungals. Of particular concern is the emergence of *C. auris*, the first fungal pathogen classified by the Centers for Disease Control and Prevention (CDC) as an urgent public health threat due to its association with extremely high mortality rates and the potential to develop resistance to all current antifungals.

Shorr A et al. [49] and Vardakas KZ et al. [50] conducted two meta-analyses of randomized clinical trials assessing prophylaxis with azoles in surgical ICU patients. These analyses concluded that there was a reduction in the incidence of fungal infections, although there was no impact on mortality. Conversely, a meta-analysis encompassing 12 studies conducted on non-neutropenic, critically ill patients determined that azole prophylaxis led to a reduction in mortality (RR, 0.76; 95% CI, 0.59–0.97) and the incidence of invasive fungal infection (RR, 0.46; 95% CI, 0.31–0.68) [49,50].

At this stage in the review, we may feel uncertain, given the observed association between the administration of antibiotics and the presence of *Candida* spp., as well as the previously documented link between prior antibiotic use and higher mortality, particularly in cases of IAC. Consequently, the question arises as to whether it is advisable to administer antibiotics to patients with peritonitis. The answer is clear: in peritonitis, antibiotics should be administered, but we must bear in mind that this creates a predisposing condition for an added or coexisting fungal infection. Consequently, we will undertake an active search for fungal infection in these patients, utilizing the available resources to initiate treatment promptly. However, we must also consider that a strategy of generalized early treatment can lead to the excessive use of antifungal medication and a possible increase in resistance to azoles or to infections caused by non-albicans species.

6. Treatment

The management of complicated intra-abdominal infection is a complex scenario where other conditions, beyond antifungal treatment, play a significant role. The patient's specific conditions, as well as early and proper control of the source of infection, are key factors in determining the outcome for these patients. Once the source of infection has been addressed, the role of antifungal therapy becomes crucial.

Several authors have used different definitions for treatments related to possible or probable *Candida* spp. infection in their studies. This variability complicates comparisons and limits the strength of conclusions. Therefore, in this review, we will apply the definitions presented in the 2019 guidelines of the European Society of Intensive Care Medicine/European Society of Infections working group [51]:

(1) "Prophylaxis" is antifungal treatment given without clear evidence of infection to critically ill patients at high risk of invasive candidiasis due to intrinsic factors (e.g., immunosuppression) or hospitalization-related factors (e.g., septic shock, abdominal surgery, prolonged ICU stay or broad-spectrum antibiotics);

(2) "Pre-emptive therapy" is antifungal treatment of critically ill patients at high risk of invasive candidiasis guided by fungal biomarkers or risk assessment scores associated with multifocal colonization;

(3) "Empirical therapy" is antifungal treatment of critically ill patients with signs and symptoms of infection (e.g., persistent fever of unknown origin) who have specific risk factors for invasive candidiasis, regardless of biomarker results;
(4) "Targeted therapy" is antifungal treatment based on microbiological findings confirming the presence of *Candida* spp. and its susceptibility to different antifungal agents.

To ensure optimal treatment efficacy and minimize the risk of re-resistance, it is essential that the antifungal agent reach adequate concentrations at the site of infection, which in this case could be intraperitoneal. It is important to emphasize that there is limited and often conflicting information on intra-abdominal antifungal concentrations in critically ill patients with peritonitis [52]. It is important to recognize the limitations of extrapolating data from blood concentrations to intraperitoneal levels. The central compartment, which includes the bloodstream, provides a convenient method for monitoring drug levels and assessing pharmacokinetic/pharmacodynamic (PK/PD) behavior to optimize drug efficacy.

However, it is well established that patients in critical condition exhibit significant variability in drug PK behavior due to several factors. These include a marked increase in drug distribution volume and elevated renal clearance, particularly during the initial phases of critical illness [53]. These PK alterations primarily impact hydrophilic drugs, such as echinocandins, which are the first-line treatment recommended by various international guidelines. If the dosage of these drugs is not properly adjusted, often requiring dose escalation or loading doses, the resulting drug concentrations may fall below the levels needed to act against pathogens with reduced susceptibility, thus promoting the emergence of so-called "resistant mutants" [54]. In cases of intra-abdominal infections, the diffusion of drugs into the peritoneal cavity is further hindered by poor penetration, as well as the presence of surgical drains (often high-output or frequently flushed). Additional factors that can impact the ability to achieve adequate drug concentrations include high body weight (over 100 kg), low albumin levels common among critically ill patients and the frequent use of intermittent or continuous renal replacement therapies [55].

It is essential to consider all these factors when selecting the most appropriate antifungal treatment for each patient's specific condition.

Three classes of drugs (azoles, echinocandins and polyenes) make up the available therapeutic arsenal for treating candidiasis and IAC. As previously mentioned, *Candida albicans* remains the most frequently isolated species. However, the frequent presence of non-albicans species and the emergence of resistance to azoles and echinocandins make empirical treatment a significant challenge that must be addressed on a daily basis (see Table 5).

Table 5. In vitro susceptibility of available antifungals (++: adequate effect; -/+: possible resistance; --: resistance; L-AMB: liposomal amphotericin B).

Fungal Species	L-AMB	Fluconazole	Voriconazole	Isavuconazole	Echinocandins
Candida albicans	++	++	++	++	++
Candida parapsilosis	++	++	++	++	+
Candida tropicalis	++	++	++	++	++
Candida glabrata	++	-/+	-/+	-/+	-/+
Candida krusei	++	--	--	--	++
Candida auris	+	--	--	++	+

6.1. Azoles

Azoles are a class of antifungal agents that disrupt ergosterol biosynthesis in the fungal membrane by inhibiting 14-α-sterol demethylase, a microsomal cytochrome (CYP). These compounds are categorized into two main groups: imidazoles (e.g., clotrimazole, miconazole and ketoconazole) and triazoles (e.g., fluconazole, voriconazole and isavuconazole). Systemic triazoles are metabolized more slowly and have fewer effects on human sterol synthesis compared to imidazoles. All azoles undergo hepatic metabolism through different cytochrome P450 (CYP) enzymes, albeit with varying intensities, leading to numerous drug interactions that must be carefully considered.

6.1.1. Fluconazole

The primary mechanism of action of fluconazole consists of the inhibition of fungal cytochrome P-450 by blocking 14-alpha-lanosterol demethylation, a key step in the biosynthesis of fungal ergosterol, and a moderate inhibitor of CYP3A4. A direct relationship has been observed between the dose, the MIC (minimum inhibitory concentration) and clinical efficacy.

In addition to multiple observed drug interactions, there is a risk of increasing plasma concentrations of other compounds metabolized by CYP2C9, CYP2C19 and CYP3A4 when co-administered with fluconazole (such as anticoagulants, antidepressants, benzodiazepines, carbamazepine, calcium channel blockers, fentanyl, sirolimus, tacrolimus, losartan, phenytoin, prednisone, etc.). Due to its long half-life, fluconazole continues to inhibit enzymes for four to five days after discontinuation [56]. Fluconazole can be administered orally (with good bioavailability, though not widely recommended in critically ill patients) or intravenously. It is important to note that food intake does not affect oral absorption. Peak plasma concentrations (Cmax) are reached between 0.5 and 1.5 h post-dose and are proportional to the administered dose. Steady-state levels are achieved within 4 to 5 days after multiple once-daily doses. A loading dose of 200 mg on day 1, which is twice the usual daily dose, increases plasma levels to 90% of steady-state concentrations by day 2, making this a recommended strategy for critically ill patients. The volume of distribution (VD) is 40 L, with low protein binding (10%). This means that while a large proportion of the drug remains free to exert its action, its renal elimination is significantly increased in hyper-filtering patients or those undergoing renal replacement therapies, requiring reconsideration of its use in such patients. Fluconazole distributes widely, achieving high concentrations in cerebrospinal fluid (CSF), the lungs, skin, the cornea and urine. The recommended loading dose is 800 mg (400 mg every 12 h) on the first day, followed by 400 mg/day. Maintenance doses may vary depending on patient conditions [56,57].

6.1.2. Voriconazole

The primary mechanism of action of voriconazole is the inhibition of 14-alpha-lanosterol demethylation, which is mediated by fungal cytochrome P-450. This is a key step in the process of fungal ergosterol biosynthesis. The accumulation of 14-alpha-methyl sterols is associated with the subsequent loss of ergosterol in the fungal cell membrane [58]. Voriconazole can be administered orally (though rarely in critically ill patients) or intravenously. Its pharmacokinetics are non-linear due to metabolic saturation, meaning that a disproportionate rise in plasma concentrations is seen with increases in dose. Administering a loading dose helps to achieve plasma concentrations close to the steady state within the first 24 h, making this a particularly important strategy for critically ill patients. The oral bioavailability of voriconazole is 96%; however, it is influenced by CYP3A4 activity and a transporter protein, as well as by high-fat meals, which reduce Cmax and AUC (area under the curve). Voriconazole distributes widely, with a volume of distribution of 4.6 L/kg. Voriconazole has a plasma protein binding of 58%, and it reaches effective concentrations in the CSF, lungs, brain, liver, kidneys and spleen. As previously mentioned, voriconazole is metabolized by, and also inhibits, cytochrome P450 enzymes CYP2C19,

has been associated with a lower AUC0-24 in critically ill patients compared to healthy volunteers. Due to its extensive protein binding, only the free fraction is considered capable of diffusing into the peritoneum, potentially compromising its efficacy. While different studies suggest that peritoneal concentrations would meet the classical efficacy targets defined by an AUC/MIC of 3000 for all *Candida* species, a recent study by Gioia F et al. [60] showed that the AUC/3000 results in serum were 0.042, 0.032 and 0.022 for anidulafungin, micafungin and caspofungin, respectively. These levels would be adequate for the treatment of *C. albicans* (0.03 mg/L) but suboptimal for *C. glabrata*, *C. krusei* and *C. tropicalis* (0.06 mg/L), potentially creating an environment conducive to resistance development.

It has been hypothesized by several authors that IAC may act as a hidden reservoir for echinocandin-resistant *C. glabrata* species. *C. glabrata*, which is increasingly present in IAC, exhibits a significant and varied adaptive response to its environment, linked to intrinsic mutations in FKS genes, which encode β-1,3-D-glucan synthase. These mutations result in an increased chitin content in the cell wall and paradoxical growth when high doses of echinocandins are administered. Finally, concentrations <2 mg/L lead to the selection of resistant mutants, which emerged rapidly (<48 h) in laboratory studies at different echinocandin concentrations. Metabolism occurs in the plasma, with no evidence of hepatic metabolism, making anidulafungin a drug with few interactions. It undergoes slow chemical degradation at physiological temperature and pH, resulting in an open-ring peptide that lacks antifungal activity. Renal elimination is minimal (<10%). A single 200 mg loading dose should be administered on day 1, followed by a daily maintenance dose of 100 mg [57,60–62].

6.2.2. Rezafungin

Rezafungin is a novel echinocandin that selectively inhibits fungal 1,3-β-D-glucan synthase with inhibition of fungal cell wall synthesis, resulting in rapid and concentration-dependent fungicidal activity in Candida species. It has state-of-the-art pharmacokinetics with remarkable stability, high solubility and an exceptionally long half-life, allowing early exposure to high drug levels. It can therefore be administered intravenously once a week, with a loading dose of 400 mg on day 1, followed by 200 mg on day 8 and once a week thereafter [63]. The pharmacokinetic results in the single ascending dose study were consistent with the post-first-dose pharmacokinetic results in the multiple ascending dose study for each dose cohort. At all doses investigated in the single ascending dose study, mean maximum plasma concentration (Cmax) values ranged from 2.76 to 22.7 g/mL, and the values corresponding to the mean area under the concentration–time curve from time zero to 168 h (AUC0-168) ranged from 145 to 1160 g-h/mL. Both Cmax and AUC increased in a dose-proportional manner, and mean half-life (t1/2) values of 80 h were observed during the first week (up to day 7) of plasma sampling (a longer terminal t1/2 of 125 to 146 h was calculated by incorporating data from later sampling times [days 14 and 21]) [64].

Rezafungin is rapidly distributed with a volume of distribution similar to that of body water (≈40 L) and binds to 97% of plasma proteins. Like the other echinocandins, it is not affected by intermittent/continuous renal replacement techniques. Furthermore, it undergoes little or no biotransformation and is mainly excreted in the feces, with less than 1% of the drug being excreted in the urine, with an average plasma half-life of 127 to 146 h [65].

Rezafungin has a broad spectrum of activity against *Candida* and *Aspergillus* species, including *Candida auris* and subsets of fungal strains resistant to other antifungals (echinocandins and azoles), and has demonstrated therapeutic and prophylactic efficacy in animal models of candidiasis, aspergillosis and Pneumocystis pneumonia. Rezafungin has demonstrated safety and tolerability comparable to current echinocandins. In addition, rezafungin has demonstrated chemical and metabolic stability and non-hepatotoxicity compared to anidulafungin [61–63], with increased biofilm activity.

In phase III trials, rezafungin has demonstrated non-inferiority to caspofungin for the treatment of invasive candidiasis (ReSTORE trial) [66]. In addition to its weekly dosing, rezafungin offers other potential pharmacokinetic and pharmacodynamic advantages, such as excellent penetration into anatomically challenging sites, such as intra-abdominal collections, and a potentially lower risk of promoting local resistance. This positions it as a promising option for deep infections, particularly intra-abdominal infections, justifying its commercialization and inclusion as a first-line agent in recent guidelines for the treatment of candidiasis [67].

6.3. Polyenes

Amphotericin B (AMB)

Liposomal amphotericin B (L-AMB) is a heptane-derived macrolide that contains seven conjugated trans double bonds and 3-amino-3-6-dideoxy-mannose (mycosamine) linked to the main ring via a glycosidic bond. The amphoteric behavior that gives the drug its name is due to the presence of a carboxyl group on the main ring and a primary amino group on mycosamine, which confer aqueous solubility at pH extremes. Its effectiveness as a fungicide or fungistatic depends on the achieved concentration and the specific fungal species. Its mechanism of action involves binding to ergosterol in the fungal membrane, leading to membrane disruption and increased permeability. Its antifungal activity is concentration-dependent, meaning it is related to the plasma concentration (Cmax) over the minimum inhibitory concentration (MIC), demonstrating a prolonged post-antifungal effect [68,69].

L-AMB achieves higher plasma concentrations (Cmax) than AMB and has a longer exposure time (AUC0-24). Following administration, L-AMB reaches the steady state rapidly, generally within the first four days of the same dose. A key feature of L-AMB is its action against biofilm formation, which is directly implicated in IAC [68,70,71]. The pharmacokinetics of L-AMB are largely dependent on the composition and size of the liposomal or lipid complex particles. Large structures, such as the lipid complex (Abelcet), are rapidly absorbed by the mononuclear phagocyte system, whereas smaller liposomes remain in the circulation for extended periods.

The main limitations to the use of AMB are nephrotoxicity and hepatotoxicity. However, AMB has been withdrawn from the market in many European countries, and lipid formulations, especially liposomal (L-AMB), are the most widely used. AMB molecules are stabilized by phospholipids and cholesterol within the liposomal bilayer, which reduces toxicity to animal cells. The small size of the liposomal particles (100 nm or less) ensures that drug levels remain high in the bloodstream, reducing distribution to organs, including the kidneys, and contributing to improved safety and reduced nephrotoxicity [62–65]. Studies by our group with real-world data have confirmed the adequate safety profile of L-AMB in critically ill patients [72].

The enhanced vascular permeability in critically ill patients has been shown to increase the transfer of L-AMB from the circulation to the lesions, thereby boosting its antifungal activity. In plasma, AMB remains associated with liposomes (97% at 4 h, 55% at 168 h) after L-AMB administration. It is assumed that L-AMB does not directly bind to target cells but instead facilitates AMB transfer from the liposomal membrane to the fungal cell membrane, which is critical for antifungal activity. The usual dose is 3–5 mg/kg/day, achieving adequate concentrations in the lungs, ascitic fluid and muscle tissue. While the drug does not penetrate the blood–brain barrier (BBB) efficiently, it is indicated for use in cases of inflammation, potentially in combination with other antifungals. Finally, due to its pharmacokinetics, plasma level monitoring is not required, as its efficacy does not correlate with observed blood levels [68–71].

7. Antifungal Therapeutic Drug Monitoring (TDM) Requirements

The importance of therapeutic drug monitoring (TDM) of antifungal agents in a variety of clinical settings is increasingly being recognized. However, there are no definitive data from large-scale clinical trials to support its use in all clinical settings (and there probably never will be). Most of the evidence supporting TDM is circumstantial. In addition, antifungal TDM can be costly and time-consuming, and its ultimate impact on clinical care may be difficult to estimate. Therefore, there is a broad consensus that TDM should not be used routinely (as is the case for some antimicrobials, such as aminoglycosides) but more selectively.

Beyond its indication, TDM is not available in most hospitals. Furthermore, TDM needs to have an adequate response time for results if treatment is to be adjusted early. Finally, most authors indicate that TDM should be determined when the antifungal is in the stable phase, which is achieved between 3 and 5 days. Detection of suboptimal levels after 72–96 h may not be useful in critically ill patients [73–75].

At this time, there is no evidence or indication to support the routine use of TDM for L-AMB, echinocandins, fluconazole and isavuconazole. Meanwhile, most authors assume that TDM is necessary for voriconazole [73–77]. However, some considerations must be taken into account.

For L-AMB, the structure of the liposomes deserves special attention. As with other liposomal formulations, drugs sequestered within this particle cannot reach diffusion equilibrium with the extravascular compartment. In addition, amphotericin B (AMB) released from a liposome binds strongly to plasma proteins (>90%, depending largely on the clinical status of the patient), and this aspect may affect the final exposure of AMB in the blood. Therefore, total AMB measured in the blood after administration of L-AMB may not be indicative of true exposure, and the clinical utility of monitoring L-AMB blood concentrations may be questionable since plasma levels are not related to clinical efficacy [75–77].

In general, it is not necessary to monitor fluconazole levels due to its favorable pharmacokinetic characteristics with rapid absorption and high bioavailability, prolonged distribution in the body and relatively high plasma levels. In addition, there is a direct correlation between the dose of fluconazole and the plasma levels achieved, which gives the drug predictable pharmacokinetics, and therefore it is not considered necessary to monitor plasma levels. However, it may be necessary to monitor plasma levels of fluconazole in certain groups of patients, in particular, patients on continuous renal replacement therapy (CRRT) or intermittent renal replacement therapy (IRRT). Although increasing the dose (800–1200 mg/day) may be an option to consider in this subgroup of patients, we believe that patients requiring CRRT/RRIT should be treated with therapy-unaffected antifungals such as L-AMB or echinocandins [73–77].

Echinocandins show a significant post-antifungal effect and therefore concentration-dependent activity. The Cmax/MIC and AUC0_24/MIC ratios (measured as total drug concentrations) are considered relevant pharmacodynamic indices for these drugs. Most experts believe that the data on the relationship between blood concentrations of echinocandins and therapeutic outcome are insufficient to support the routine use of DMT for these agents. However, factors such as obesity, age and clinical status may influence exposure and contribute to significant pharmacokinetic differences between these drugs. Although this PK/PD is described as a guarantee of efficacy for the treatment of *Candida* spp. infections and the levels far exceed the MIC90 for common *Candida* spp. strains, they would be insufficient for the treatment of *C. parapsilosis* or *C. glabrata* in intra-abdominal infections. Therefore, if TDM is not possible, we believe that loading doses and high doses of echinocandins should be used in critically ill patients [73–77].

Isavuconazole has linear and dose-proportional pharmacokinetics, which is useful for predicting blood concentrations in humans. It also has lower rates of adverse events and a better safety profile than other triazoles, with a lower propensity for cytochrome P450-mediated drug–drug interactions than voriconazole. Most experts agree that TDM is not necessary in patients receiving isavuconazole [73–77].

Most researchers maintain that TDM should be performed routinely in most patients receiving voriconazole. This azole has highly variable intra- and interindividual pharmacokinetics, attributed to different factors, such as pharmacogenetic polymorphisms, drug interactions (remember that it is mainly metabolized by CYP2C19, 2C9 and 3A4), alterations in gastrointestinal absorption, and even inflammation and the severity of the patient's condition, as well as body weight. In our opinion, if TDM is not possible, voriconazole is not a treatment option for critical patients [73–77].

8. Combination Therapy Requirements

In general, the antifungal combination is not frequently used for invasive candidiasis or IAC, except possibly for *C. auris* because of its high level of resistance [78]. The international guidelines for the management of candidiasis generally do not contemplate a combination therapy except in some clinical circumstances such as CNS infections and endocarditis in which 5 flucytosine (5-FC) can be added to AMB [78–80].

Ideally, the overall goal of combination antifungal therapy is to achieve increased clinical efficacy of the antifungal agent and avoid toxicity to the patient. There are several advantages and reasons to consider a combination of drugs: (i) potential increased potency and extent of fungal killing (synergy); (ii) a broader spectrum of activity targeting potentially resistant pathogens; (iii) prevention of microbial resistance; and (iv) combination therapy may allow reduced doses of individual antifungal drugs, which may minimize toxicities.

It is equally important to consider potential antagonistic mechanisms, particularly with regard to the interaction between azoles and amphotericin B. Azoles bind avidly to ergosterol, causing its depletion in cytoplasmic membranes. Therefore, in combination with AMB, there are fewer binding sites available for AMB, resulting in a decrease in AMB binding and antifungal activity. However, the intensity of this interaction depends on the PK characteristics of the azoles. For example, highly lipophilic azoles (itraconazole) that accumulate in the cell membrane reduce the effect of AMB. On the other hand, hydrophilic azoles (fluconazole) do not accumulate in the membrane and therefore do not show antagonism [77]. In addition, there may be another type of antagonism caused by pre-exposure to azoles, which causes changes in the ergosterol in the membrane and reduces the effectiveness of AMB. In addition, there may be another type of antagonism caused by pre-exposure to azoles, which causes changes in the ergosterol in the membrane and reduces the effectiveness of AMB [78–80].

Although the combination of echinocandins with fluconazole has achieved a good response, no differences have been shown with respect to treatment with fluconazole alone. On the other hand, the combination of AMB with echinocandins exhibits erratic behavior ranging from a few cases of synergism to no action being evidenced or antagonism appearing in the majority of cases [79,80].

5-Flucytosine (5-FC) has a unique mechanism of action that theoretically makes it an ideal add-on drug for use in combination therapies, as it acts on RNA and prevents protein synthesis in fungi. It also inhibits DNA synthesis but is inactive against bacteria and human cells. Most in vitro studies have used this drug in combination with AMB, followed by combinations with azoles and echinocandins. Several studies have been conducted with difficult-to-treat *Candida* species. In a recent study of two-drug combinations against multidrug-resistant C. auris, 5-FC at 1.0 mg/L was found to improve most combinations with AMB, echinocandins and voriconazole [81].

However, except in specific situations (endocarditis or central nervous system infection), its administration in critically ill patients is not common because 5-FC has a well-known side-effect profile that includes both dose-related pharmacological toxicity and idiosyncratic pharmacological toxicity, including bone marrow toxicity (leukopenia and thrombocytopenia), hepatotoxicity, gastrointestinal intolerance and renal impairment, which generally requires TDM of this agent [81].

Development of New Antifungal Agents

Based on the above, research continues to identify new antifungal targets and to develop specific drugs against these targets and to prevent resistance. Studies on antifungal drug resistance have identified several effective strategies such as increasing membrane β-glucan, improving tolerance through cellular stress, inhibiting biofilm formation and suppressing ergosterol biosynthesis (ERG). In addition, targets related to enzymes (acetyltransferases and deacetylases), fungal aspartate pathways, HSP90, CYP51, HDAC, SE, fructose bisphosphate aldolase (FBA), arachidonic acid pathways and sulphite transporters have gained in importance [82–84].

Several single-target drugs, such as suba-itraconazole, VT-1129, VT-1161 and VT-1598, and cell wall targeting drugs, such as amphotericin B coheates (CAmB), ibrexafungerp and fosmanogepix, as well as drugs targeting intracellular targets, including VL-2397, T-2307, MGCD290 and olorofim, are being studied intensively [76–78]. However, toxicity remains an issue, and to overcome these limitations, many research groups have focused on molecular hybridization to create multitarget drugs rather than traditional drug combinations. This shift has led to the design of multitarget ligands that are capable of acting simultaneously on multiple sites essential to the fungal life cycle with a single molecule, producing synergistic effects. This new approach could overcome challenges such as the development of resistance, limited pharmacokinetics and poor compliance associated with single-target drugs. The development of new drugs is of paramount importance for the future of patients with invasive fungal infections, but their detailed development is beyond the scope of this review.

9. When and How to Treat? The Role of Guidelines

The first question to consider is *When should I start empirical treatment for my patient?* As previously discussed, the decision to initiate prophylactic, anticipatory or empirical treatment will depend on the clinical conditions and risk factors of the patient. While most authors advocate against the routine use of antifungal prophylaxis in critically ill patients, we generally support the initiation of prophylactic treatment in patients with peritonitis and perforation of the upper gastrointestinal tract (above the Treitz angle), particularly in patients who use antacids and have gastric neoplasms. Conversely, in patients with risk factors and positive biomarkers (BDG), we may initiate anticipatory treatment, though this approach is subject to greater controversy, particularly regarding the cost-effectiveness of biomarkers for diagnosis. Finally, in high-risk patients with fever or shock and an unclear diagnosis, initiating empirical treatment may be a consideration.

The second question pertains to *the most appropriate treatment approach*. As outlined in the objectives, the purpose of this review is to provide readers with the tools to understand different aspects of invasive candidiasis and IAC, as well as the PK/PD characteristics of various antifungals. This will empower each physician to develop an empirical treatment protocol for patients at risk of invasive candidiasis/IAC tailored to the local situation and available resources. However, it is essential to review the international guidelines, which aim to serve as a basis for our local adaptation of treatment. As outlined in Table 6, the main guidelines provide treatment recommendations. Echinocandins are the most frequently mentioned drugs as first-line treatment for invasive candidiasis, and no specific recommendations are made for IAC, with the same treatment being accepted. However, as discussed previously, especially in cases where *C. glabrata* is suspected, this may not be the most suitable option. A recent systematic review has concluded that there are no significant differences in clinical efficacy between treatment with L-AMB, echinocandins or voriconazole in critically ill patients with invasive candidiasis. Therefore, the authors suggest that these three types of drugs should be considered for first-line treatment [1,3,4,51,57,67], and the definitive drug should be chosen based on the isolated species and the susceptibility profile. The use of L-AMB as first-line treatment is supported by various authors, based on the low likelihood of resistance development, especially in patients with a history of prior azole use. Conversely, fluconazole may not be the optimal choice for empirical treatment in patients at risk of invasive candidiasis or IAC due to the high frequency of resistance observed in non-albicans species. Therefore, as is standard practice with antibiotic treatment, an empirical broad-spectrum treatment should be initiated and then de-escalated once microbiological results confirm the species' susceptibility to fluconazole.

Table 6. Summary of recommendations from major international guidelines for empirical treatment in non-neutropenic patients with invasive candidiasis and IAC. (Recommendation A = Strong recommendation for use; B = Moderate recommendation; C = Marginal recommendation; I = High-quality evidence; II = Moderate-quality evidence; III = Expert opinion.)

Invasive Candidiasis	ECIL Recommendation	IDSA Recommendation	ESCMID Recommendation	ESICM/ESCMID Recommendation	ECMM Recommendation
General population	Candins/AI	Candins/strong	Candins/AI	Candins	Candins/AI
	L-AMB/AI	L-AMB/strong	L-AMB/BI	L-AMB as rescue	L-AMB/AI
	Fluconazole/AI	Candins/strong (non-critical patients)	Fluconazole/CI	Fluconazole in patients without DMO	
	Voriconazole/AI	Voriconazole/strong	Voriconazole/BI	Not considered	
IAC	Non-special considerations	Non-special considerations	Non-special considerations		Non-special considerations

ECIL: The European Conference on Infections in Leukemia [85]; IDSA: Infectious Diseases Society of America [57], ESCMID: European Society of Clinical Microbiology and Infectious Diseases [86], ESICM: The European Society of Intensive Care Medicine [51], ECMM: European Confederation of Medical Mycology [67].

Finally, Figure 1 proposes an algorithm for the treatment of patients with ICA that can serve as a basis for the development of a local algorithm, depending on the epidemiology, the particular resources of each unit and the species of *Candida* isolated.

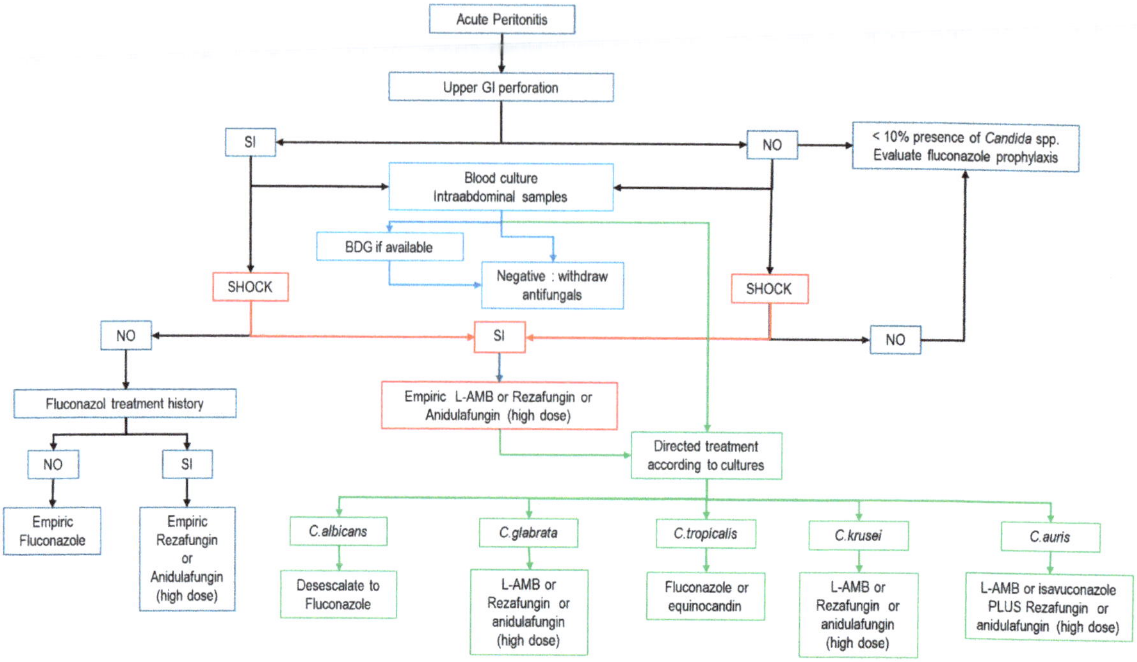

Figure 1. Empirical treatment algorithm for IAC. (BDG: beta-d-glucan; L-AMB: liposomal amphotericin B; modified from refs. [1,51,57,67,85,86].) Blue arrows: possible action according to biomarkers; Red arrows: possible action according to the presence of shock; Green arrows: possible action for targeted therapy.

10. Conclusions

Candidemia and intra-abdominal candidiasis are relatively common complications in critically ill patients associated with high morbidity and mortality. The challenges posed by increasing resistance to antifungal drugs, the need for new and better diagnostic tools, and the development of personalized treatment strategies could improve the grim prognosis of this condition. Therefore, research into new antifungal drugs with different targets and biomarkers that allow more accurate early diagnosis is needed, as well as the development of local and national strategies to control resistance.

Author Contributions: Conceptualization, M.M.P., A.R., R.Z., F.F.G.-B. and I.M.-L.; methodology, M.M.P., A.R. and F.F.G.-B.; investigation, M.M.P., R.Z., I.M.-L., F.F.G.-B. and A.R.; writing—original draft preparation, M.M.P. and A.R.; writing—review and editing, M.M.P., R.Z., I.M.-L., F.F.G.-B. and A.R.; funding acquisition, A.R. All authors have read and agreed to the published version of the manuscript.

Funding: This study was supported with protected research time (AR) by a grant from the Ricardo Barri Casanovas Foundation (FRBC01/2024). The sponsors were not involved in the study design, data collection, data analysis, data interpretation or report writing.

Conflicts of Interest: The authors declare no conflicts of interest.

Abbreviations

The following abbreviations are used in this manuscript:

IAC	Intra-abdominal candidiasis
L-AMB	Liposomal amphotericin B
AMB	Amphotericin B
PK/PD	Pharmacokinetic/pharmacodynamic
TDM	Therapeutic drug monitoring
AUC	Area under the curve
MIC	Minimum inhibitory concentration
CYP	Cytochrome
Cmax	Peak plasma concentration
VD	Volume of distribution

References

1. Maseda, E.; Martín-Loeches, I.; Zaragoza, R.; Pemán, J.; Fortún, J.; Grau, S.; Aguilar, G.; Varela, M.; Borges, M.; Giménez, M.-J.; et al. Critical appraisal beyond clinical guidelines for intraabdominal candidiasis. *Crit. Care* **2023**, *27*, 382. [CrossRef] [PubMed] [PubMed Central]
2. Bassetti, M.; Giacobbe, D.R.; Vena, A.; Trucchi, C.; Ansaldi, F.; Antonelli, M.; Adamkova, V.; Alicino, C.; Almyroudi, M.-P.; Atchade, E.; et al. Incidence and outcome of invasive candidiasis in intensive care units (ICUs) in Europe: Results of the EUCANDICU project. *Crit. Care* **2019**, *23*, 219. [CrossRef]
3. Zaragoza, R.; Ramírez, P.; Borges, M.; Pemán, J. Puesta al día en la candidiasis invasora en el paciente crítico no neutropénico [Update on invasive candidiasis in non-neutropenic critically ill adult patients]. *Rev. Iberoam. Micol.* **2016**, *33*, 145–151. [CrossRef]
4. Candel, F.J.; Pacheco, C.P.; Ruiz-Camps, I.; Maseda, E.; Sánchez-Benito, M.R.; Montero, A.; Puig, M.; Gilsanz, F.; Aguilar, J.; Matesanz, M. Update on management of invasive candidiasis. *Rev. Esp. Quimioter.* **2017**, *30*, 397–406.
5. Azim, A.; Ahmed, A. Diagnosis and management of invasive fungal diseases in non-neutropenic ICU patients, with focus on candidiasis and aspergillosis: A comprehensive review. *Front. Cell. Infect. Microbiol.* **2024**, *14*, 1256158. [CrossRef]
6. van Ruler, O.; Boermeester, M.A. Surgical treatment of secondary peritonitis: A continuing problem. *Chirurg* **2017**, *88* (Suppl. S1), 1–6. [CrossRef]
7. van Ruler, O.; Mahler, C.W.; Boer, K.R.; Reuland, E.A.; Gooszen, H.G.; Opmeer, B.C.; de Graaf, P.W.; Lamme, B.; Gerhards, M.F.; Steller, E.P.; et al. Comparison of On-Demand vs Planned Relaparotomy Strategy in Patients With Severe Peritonitis. A Randomized Trial. *JAMA* **2007**, *298*, 865–872. [CrossRef]
8. Schroeder, J.A.M.; Wilson, C.M.; Pappas, P.G. Invasive Candidiasis. *Infect. Dis. Clin. N. Am.* **2025**, *39*, 93–119. [CrossRef] [PubMed]
9. Pfaller, M.A.; Diekema, D.J. Epidemiology of invasive candidiasis: A persistent public health problem. *Clin. Microbiol. Rev.* **2007**, *20*, 133–163. [CrossRef] [PubMed]
10. Lockhart, S.R. Current Epidemiology of Candida Infection. *Clin. Microbiol. Newsl.* **2014**, *36*, 131–136. [CrossRef] [PubMed]
11. Chen, X.; Zhang, R.; Takada, A.; Iwatani, S.; Oka, C.; Kitamoto, T.; Kajiwara, S. The role of Bgl2p in the transition to filamentous cells during biofilm formation by Candida albicans. *Mycoses* **2017**, *60*, 96–103. [CrossRef] [PubMed]
12. García, C.S.; Palop, N.T.; Bayona, J.V.M.; García, M.M.; Rodríguez, D.N.; Álvarez, M.B.; Serrano, M.D.R.G.; Cardona, C.G. Candida auris: Report of an outbreak. *Enferm. Infecc. Microbiol. Clin.* **2020**, *38* (Suppl. S1), 39–44. [CrossRef]
13. Chowdhary, A.; Jain, K.; Chauhan, N. Candida auris Genetics and Emergence. *Annu. Rev. Microbiol.* **2023**, *77*, 583–602. [CrossRef] [PubMed]
14. Bing, J.; Guan, Z.; Zheng, T.; Ennis, C.L.; Nobile, C.J.; Chen, C.; Chu, H.; Huang, G. Rapid evolution of an adaptive multicellular morphology of Candida auris during systemic infection. *Nat. Commun.* **2024**, *15*, 2381. [CrossRef]
15. Kullberg, B.J.; Arendrup, M.C. Invasive Candidiasis. *N. Engl. J. Med.* **2015**, *373*, 1445–1456. [CrossRef]
16. Lamoth, F.; Lockhart, S.R.; Berkow, E.L.; Calandra, T. Changes in the epidemiological landscape of invasive candidiasis. *J. Antimicrob. Chemother.* **2018**, *73*, i4–i13. [CrossRef] [PubMed]
17. Montoya, M.C.; Moye-Rowley, W.S.; Krysan, D.J. Candida auris: The canary in the mine of antifungal drug resistance. *ACS Infect. Dis.* **2019**, *5*, 1487–1492. [CrossRef]
18. Díaz-García, J.; Machado, M.; Alcalá, L.; Reigadas, E.; Pérez-Ayala, A.; de la Pedrosa, E.G.-G.; Gónzalez-Romo, F.; Cuétara, M.S.; García-Esteban, C.; Quiles-Melero, I.; et al. Trends in antifungal resistance in Candida from a multicenter study conducted in Madrid (CANDIMAD study): Fluconazole-resistant C. parapsilosis spreading has gained traction in 2022. *Antimicrob. Agents Chemother.* **2023**, *67*, e0098623. [CrossRef]

19. Arendrup, M.C.; Guinea, J.; Meletiadis, J. Twenty Years in EUCAST Anti-Fungal Susceptibility Testing: Progress & Remaining Challenges. *Mycopathologia* **2024**, *189*, 64. [CrossRef]
20. Dupont, H.; Paugam-Burtz, C.; Muller-Serieys, C.; Fierobe, L.; Chosidow, D.; Marmuse, J.-P.; Mantz, J.; Desmonts, J.-M. Predictive factors of mortality due to polymicrobial peritonitis with Candida isolation in peritoneal fluid in critically ill patients. *Arch. Surg.* **2002**, *137*, 1341–1346. [CrossRef]
21. Vergidis, P.; Clancy, C.J.; Shields, R.K.; Park, S.Y.; Wildfeuer, B.N.; Simmons, R.L.; Nguyen, M.H. Intra-Abdominal Candidiasis: The Importance of Early Source Control and Antifungal Treatment. *PLoS ONE* **2016**, *11*, e0153247. [CrossRef]
22. ENVIN-HELICS: Estudio Nacional de Vigilancia de Infección Nosocomial en Servicios de Medicina Intensiva. Available online: https://hws.vhebron.net/envin-helics/ (accessed on 24 April 2025).
23. Bassetti, M.; Vena, A.; Giacobbe, D.R.; Trucchi, C.; Ansaldi, F.; Antonelli, M.; Adamkova, V.; Alicino, C.; Almyroudi, M.-P.; Atchade, E.; et al. Risk Factors for Intra-Abdominal Candidiasis in Intensive Care Units: Results from EUCANDICU Study. *Infect. Dis. Ther.* **2022**, *11*, 827–840. [CrossRef]
24. Bassetti, M.; Vena, A.; Bouza, E.; Peghin, M.; Muñoz, P.; Righi, E.; Pea, F.; Lackner, M.; Lass-Flörl, C. Antifungal susceptibility testing in Candida, Aspergillus and Cryptococcus infections: Are the MICs useful for clinicians? *Clin. Microbiol. Infect.* **2020**, *26*, 1024–1033. [CrossRef] [PubMed]
25. Guarner, J.; Brandt, M.E. Histopathologic diagnosis of fungal infections in the 21st century. *Clin. Microbiol. Rev.* **2011**, *24*, 247–280. [CrossRef] [PubMed]
26. Rawson, T.M.; Antcliffe, D.B.; Wilson, R.C.; Abdolrasouli, A.; Moore, L.S.P. Management of Bacterial and Fungal Infections in the ICU: Diagnosis, Treatment, and Prevention Recommendations. *Infect. Drug Resist.* **2023**, *16*, 2709–2726. [CrossRef]
27. Wang, F.; Zhang, J.; Zhang, Q.; Song, Z.; Xin, C. Antifungal activities of Equol against Candida albicans in vitro and in vivo. *Virulence* **2024**, *15*, 2404256. [CrossRef] [PubMed]
28. Guinan, J.; Wang, S.; Hazbun, T.R.; Yadav, H.; Thangamani, S. Antibiotic-induced decreases in the levels of microbial-derived short-chain fatty acids correlate with increased gastrointestinal colonization of Candida albicans. *Sci. Rep.* **2019**, *9*, 8872. [CrossRef]
29. Barantsevich, N.; Barantsevich, E. Diagnosis and Treatment of Invasive Candidiasis. *Antibiotics* **2022**, *11*, 718. [CrossRef]
30. Soriano, A.; Honore, P.M.; Puerta-Alcalde, P.; Garcia-Vidal, C.; Pagotto, A.; Gonçalves-Bradley, D.C.; Verweij, P.E. Invasive candidiasis: Current clinical challenges and unmet needs in adult populations. *J. Antimicrob. Chemother.* **2023**, *78*, 1569–1585. [CrossRef]
31. Arvanitis, M.; Anagnostou, T.; Fuchs, B.B.; Caliendo, A.M.; Mylonakis, E. Molecular and nonmolecular diagnostic methods for invasive fungal infections. *Clin. Microbiol. Rev.* **2014**, *27*, 490–526. [CrossRef]
32. Pemán, J.; Cantón, E.; Gobernado, M.; Spanish ECMM Working Group on Candidaemia. Epidemiology and antifungal susceptibility of Candida species isolated from blood: Results of a 2-year multicentre study in Spain. *Eur. J. Clin. Microbiol. Infect. Dis.* **2005**, *24*, 23–30. [CrossRef] [PubMed]
33. Clancy, C.J.; Nguyen, M.H. Diagnosing Invasive Candidiasis. *J. Clin. Microbiol.* **2018**, *56*, e01909-17. [CrossRef] [PubMed]
34. Hall, L.; Le Febre, K.M.; Deml, S.M.; Wohlfiel, S.L.; Wengenack, N.L. Evaluation of the yeast traffic light PNA FISH probes for identification of Candida species from positive blood cultures. *J. Clin. Microbiol.* **2012**, *50*, 1446–1448. [CrossRef]
35. Delavy, M.; Dos Santos, A.R.; Heiman, C.M.; Coste, A.T. Investigating Antifungal Susceptibility in Candida Species With MALDI-TOF MS-Based Assays. *Front. Cell. Infect. Microbiol.* **2019**, *9*, 19. [CrossRef]
36. Steuber, T.D.; Butler, L.; Sawyer, A.; Chappell, R.; Edwards, J. Comparison of blood cultures versus T2 Candida Panel in management of candidemia at a large community hospital. *Eur. J. Clin. Microbiol. Infect. Dis.* **2021**, *40*, 997–1001. [CrossRef]
37. León, C.; Ruiz-Santana, S.; Saavedra, P.; Castro, C.; Loza, A.; Zakariya, I.; Úbeda, A.; Parra, M.; Macías, D.; Tomás, J.I.; et al. Contribution of Candida biomarkers and DNA detection for the diagnosis of invasive candidiasis in ICU patients with severe abdominal conditions. *Crit. Care* **2017**, *21*, 107. [CrossRef]
38. Li, S.; Xv, Y.; Sun, Y.; Shen, Z.; Hao, R.; Yan, J.; Liu, M.; Liu, Z.; Jing, T.; Li, X.; et al. Macrophage-derived CD36+ exosome subpopulations as novel biomarkers of Candida albicans infection. *Sci. Rep.* **2024**, *14*, 14723. [CrossRef] [PubMed]
39. Pittet, D.; Monod, M.; Suter, P.M.; Frenk, E.; Auckenthaler, R. Candida colonization and subsequent infections in critically ill surgical patients. *Ann. Surg.* **1994**, *220*, 751–758. [CrossRef]
40. León, C.; Ruiz-Santana, S.; Saavedra, P.; Almirante, B.; Nolla-Salas, J.; Alvarez-Lerma, F.; Garnacho-Montero, J.; León, M.A.; EPCAN Study Group. A bedside scoring system ("Candida score") for early antifungal treatment in nonneutropenic critically ill patients with Candida colonization. *Crit. Care Med.* **2006**, *34*, 730–737. [CrossRef]
41. Leroy, G.; Lambiotte, F.; Thévenin, D.; Lemaire, C.; Parmentier, E.; Devos, P.; Leroy, O. Evaluation of "Candida score" in critically ill patients: A prospective, multicenter, observational, cohort study. *Ann. Intensive Care* **2011**, *1*, 50. [CrossRef]
42. Hermsen, E.D.; Zapapas, M.K.; Maiefski, M.; Rupp, M.E.; Freifeld, A.G.; Kalil, A.C. Validation and comparison of clinical prediction rules for invasive candidiasis in intensive care unit patients: A matched case-control study. *Crit. Care* **2011**, *15*, R198. [CrossRef]

43. Salmanton-García, J.; Cornely, [illegible] Steinmann, J.; Siváková, A.; Akalin, E.H.; Arikan-Akdagli, S.; Loughlin, L.; Toscano, C.; et al. Attributable mortality of candidemia—Results from the ECMM Candid[illegible] Observational Cohort Study. *J. Infect.* **2024**, *89*, 106229. [CrossRef] [PubMed]
44. Morrell, M.; Fraser, V.J.; Kollef, M.H. Delaying the empiric treatment of candida bloodstream infection until positive blood culture results are obtained: A potential risk factor for hospital mortality. *Antimicrob. Agents Chemother.* **2005**, *49*, 3640–3645. [CrossRef] [PubMed]
45. Bassetti, M.; Merelli, M.; Ansaldi, F.; de Florentiis, D.; Sartor, A.; Scarparo, C.; Callegari, A.; Righi, E. Clinical and therapeutic aspects of candidemia: A five year single centre study. *PLoS ONE* **2015**, *10*, e0127534. [CrossRef]
46. Garey, K.W.; Rege, M.; Pai, M.P.; Mingo, D.E.; Suda, K.J.; Turpin, R.S.; Bearden, D.T. Time to initiation of fluconazole therapy impacts mortality in patients with candidemia: A multi-institutional study. *Clin. Infect. Dis.* **2006**, *43*, 25–31. [CrossRef]
47. Zilberberg, M.D.; Kollef, M.H.; Arnold, H.; Labelle, A.; Micek, S.T.; Kothari, S.; Shorr, A.F. Inappropriate empiric antifungal therapy for candidemia in the ICU and hospital resource utilization: A retrospective cohort study. *BMC Infect. Dis.* **2010**, *10*, 150. [CrossRef] [PubMed]
48. Maseda, E.; Ramírez, S.; Picatto, P.; Peláez-Peláez, E.; García-Bernedo, C.; Ojeda-Betancur, N.; Aguilar, G.; Forés, B.; Solera-Marín, J.; Aliaño-Piña, M.; et al. Critically ill patients with community-onset intraabdominal infections: Influence of healthcare exposure on resistance rates and mortality. *PLoS ONE* **2019**, *14*, e0223092. [CrossRef]
49. Shorr, A.F.; Chung, K.; Jackson, W.L.; Waterman, P.E.; Kollef, M.H. Fluconazole prophylaxis in critically ill surgical patients: A meta-analysis. *Crit. Care Med.* **2005**, *33*, 1928–1935. [CrossRef]
50. Vardakas, K.Z.; Samonis, G.; Michalopoulos, A.; Soteriades, E.S.; Falagas, M.E. Antifungal prophylaxis with azoles in high-risk, surgical intensive care unit patients: A meta-analysis of randomized, placebo-controlled trials. *Crit. Care Med.* **2006**, *34*, 1216–1224. [CrossRef]
51. Martin-Loeches, I.; Antonelli, M.; Cuenca-Estrella, M.; Dimopoulos, G.; Einav, S.; De Waele, J.J.; Garnacho-Montero, J.; Kanj, S.S.; Machado, F.R.; Montravers, P.; et al. ESICM/ESCMID task force on practical management of invasive candidiasis in critically ill patients. *Intensive Care Med.* **2019**, *45*, 789–805. [CrossRef]
52. Sartelli, M.; Catena, F.; Abu-Zidan, F.M.; Ansaloni, L.; Biffl, W.L.; Boermeester, M.A.; Ceresoli, M.; Chiara, O.; Coccolini, F.; De Waele, J.J.; et al. Management of intra-abdominal infections: Recommendations by the WSES 2016 consensus conference. *World J. Emerg. Surg.* **2017**, *12*, 22. [CrossRef] [PubMed]
53. Roger, C. Understanding antimicrobial pharmacokinetics in critically ill patients to optimize antimicrobial therapy: A narrative review. *J. Intensive Med.* **2024**, *4*, 287–298. [CrossRef]
54. Theuretzbacher, U. Pharmacokinetics/pharmacodynamics of echinocandins. *Eur. J. Clin. Microbiol. Infect. Dis.* **2004**, *23*, 805–812. [CrossRef]
55. Teixeira, J.P.; Neyra, J.A.; Tolwani, A. Continuous KRT: A Contemporary Review. *Clin. J. Am. Soc. Nephrol.* **2023**, *18*, 256–269. [CrossRef]
56. Grant, S.M.; Clissold, S.P. Fluconazole. A review of its pharmacodynamic and pharmacokinetic properties, and therapeutic potential in superficial and systemic mycoses. *Drugs* **1990**, *39*, 877–916. [CrossRef] [PubMed]
57. Pappas, P.G.; Kauffman, C.A.; Andes, D.R.; Clancy, C.J.; Marr, K.A.; Ostrosky-Zeichner, L.; Reboli, A.C.; Schuster, M.G.; Vazquez, J.A.; Walsh, T.J.; et al. Clinical Practice Guideline for the Management of Candidiasis: 2016 Update by the Infectious Diseases Society of America. *Clin. Infect. Dis.* **2016**, *62*, e1–e50. [CrossRef]
58. Greer, N.D. Voriconazole: The newest triazole antifungal agent. *Bayl. Univ. Med. Cent. Proc.* **2003**, *16*, 241–248. [CrossRef] [PubMed]
59. Lewis, J.S., 2nd; Wiederhold, N.P.; Hakki, M.; Thompson, G.R., 3rd. New Perspectives on Antimicrobial Agents: Isavuconazole. *Antimicrob. Agents Chemother.* **2022**, *66*, e0017722. [CrossRef]
60. Gioia, F.; Gomez-Lopez, A.; Alvarez, M.E.; Gomez-García de la Pedrosa, E.; Martín-Davila, P.; Cuenca-Estrella, M.; Moreno, S.; Fortun, J. Pharmacokinetics of echinocandins in suspected candida peritonitis: A potential risk for resistance. *Int. J. Infect. Dis.* **2020**, *101*, 24–28. [CrossRef]
61. Hall, R.G., 2nd; Liu, S.; Putnam, W.C.; Kallem, R.; Gumbo, T.; Pai, M.P. Optimizing anidulafungin exposure across a wide adult body size range. *Antimicrob. Agents Chemother.* **2023**, *67*, e0082023. [CrossRef]
62. Sganga, G.; Wang, M.; Capparella, M.R.; Tawadrous, M.; Yan, J.L.; Aram, J.A.; Montravers, P. Evaluation of anidulafungin in the treatment of intra-abdominal candidiasis: A pooled analysis of patient-level data from 5 prospective studies. *Eur. J. Clin. Microbiol. Infect. Dis.* **2019**, *38*, 1849–1856. [CrossRef] [PubMed]
63. Ham, Y.Y.; Lewis, J.S., 2nd; Thompson, G.R., 3rd. Rezafungin: A novel antifungal for the treatment of invasive candidiasis. *Future Microbiol.* **2021**, *16*, 27–36. [CrossRef] [PubMed]
64. Bassetti, M.; Stewart, A.; Bartalucci, C.; Vena, A.; Giacobbe, D.R.; Roberts, J. Rezafungin acetate for the treatment of candidemia and invasive candidiasis: A pharmacokinetic evaluation. *Expert Opin. Drug Metab. Toxicol.* **2025**, *21*, 125–132. [CrossRef]

65. Andes, D.; Brüggemann, R.J.; Flanagan, S.; Lepak, A.J.; Lewis, R.E.; Ong, V.; Rubino, C.M.; Sandison, T. The distinctive pharmacokinetic profile of rezafungin, a long-acting echinocandin developed in the era of modern pharmacometrics. *J. Antimicrob. Chemother.* **2025**, *80*, 18–28. [CrossRef]
66. Thompson, G.R.; Soriano, A.; Cornely, O.A.; Kullberg, B.J.; Kollef, M.; Vazquez, J.; Honore, P.M.; Bassetti, M.; Pullman, J.; Chayakulkeeree, M.; et al. Rezafungin versus caspofungin for treatment of candidaemia and invasive candidiasis (ReSTORE): A multicentre, double-blind, double-dummy, randomised phase 3 trial. *Lancet* **2023**, *401*, 49–59. [CrossRef] [PubMed]
67. Cornely, O.A.; Sprute, R.; Bassetti, M.; Chen, S.C.-A.; Groll, A.H.; Kurzai, O.; Lass-Flörl, C.; Ostrosky-Zeichner, L.; Rautemaa-Richardson, R.; Revathi, G.; et al. Global guideline for the diagnosis and management of candidiasis: An initiative of the ECMM in cooperation with ISHAM and ASM. *Lancet Infect. Dis.* **2025**, *25*, E280–E293. [CrossRef]
68. Azanza-Perea, J.R. Anfotericina B liposomal: Farmacología clínica, farmacocinética y farmacodinamia [Liposomal amphotericin B: Clinical pharmacology, pharmacokinetics and pharmacodynamics]. *Rev. Iberoam. Micol.* **2021**, *38*, 52–55. [CrossRef]
69. Beredaki, M.I.; Arendrup, M.C.; Pournaras, S.; Meletiadis, J. Comparative pharmacodynamics and dose optimization of liposomal amphotericin B against Candida species in an in vitro pharmacokinetic/pharmacodynamic model. *Antimicrob. Agents Chemother.* **2024**, *68*, e0022524. [CrossRef]
70. Miceli, M.H.; Chandrasekar, P. Safety and efficacy of liposomal amphotericin B for the empirical therapy of invasive fungal infections in immunocompromised patients. *Infect. Drug Resist.* **2012**, *5*, 9–16. [CrossRef]
71. Adler-Moore, J.P.; Gangneux, J.P.; Pappas, P.G. Comparison between liposomal formulations of amphotericin B. *Med. Mycol.* **2016**, *54*, 223–231. [CrossRef]
72. Sacanella, I.; Esteve-Pitarch, E.; Guevara-Chaux, J.; Berrueta, J.; García-Martínez, A.; Gómez, J.; Casarino, C.; Alés, F.; Canadell, L.; Martín-Loeches, I.; et al. A Real-World Data Observational Analysis of the Impact of Liposomal Amphotericin B on Renal Function Using Machine Learning in Critically Ill Patients. *Antibiotics* **2024**, *13*, 760. [CrossRef] [PubMed]
73. Ashbee, H.R.; Barnes, R.A.; Johnson, E.M.; Richardson, M.D.; Gorton, R.; Hope, W.W. Therapeutic drug monitoring (TDM) of antifungal agents: Guidelines from the British Society for Medical Mycology. *J. Antimicrob. Chemother.* **2014**, *69*, 1162–1176. [CrossRef] [PubMed]
74. Grau, S.; Luque, S. Antifungal therapeutic drug monitoring: When, how, and why. *Enferm. Infecc. Microbiol. Clin.* **2015**, *33*, 295–297. [CrossRef]
75. Hamada, Y.; Yagi, Y. Therapeutic drug monitoring of azole antifungal agents. *J. Infect. Chemother.* **2025**, *31*, 102535. [CrossRef]
76. Vena, A.; Muñoz, P.; Mateos, M.; Guinea, J.; Galar, A.; Pea, F.; Alvarez-Uria, A.; Escribano, P.; Bouza, E. Therapeutic Drug Monitoring of Antifungal Drugs: Another Tool to Improve Patient Outcome? *Infect. Dis. Ther.* **2020**, *9*, 137–149. [CrossRef]
77. Gomez-Lopez, A. Antifungal therapeutic drug monitoring: Focus on drugs without a clear recommendation. *Clin. Microbiol. Infect.* **2020**, *26*, 1481–1487. [CrossRef] [PubMed]
78. Toepfer, S.; Keniya, M.V.; Lackner, M.; Monk, B.C. Azole Combinations and Multi-Targeting Drugs That Synergistically Inhibit Candidozyma auris. *J. Fungi* **2024**, *10*, 698. [CrossRef]
79. Fioriti, S.; Brescini, L.; Pallotta, F.; Canovari, B.; Morroni, G.; Barchiesi, F. Antifungal Combinations against Candida Species: From Bench to Bedside. *J. Fungi* **2022**, *8*, 1077. [CrossRef]
80. Baddley, J.W.; Pappas, P.G. Antifungal Combination Therapy. Clinical Potential. *Drugs* **2005**, *65*, 1461–1480. [CrossRef]
81. Sigera, L.S.M.; Denning, D.W. Flucytosine and its clinical usage. *Ther. Adv. Infect. Dis.* **2023**, *10*, 1–15. [CrossRef]
82. Zobi, C.; Algul, O. The Significance of Mono- and Dual-Effective Agents in the Development of New Antifungal Strategies. *Chem. Biol. Drug Des.* **2025**, *105*, e70045. [CrossRef] [PubMed]
83. Zhang, Z.; Bills, G.F.; An, Z. Advances in the Treatment of Invasive Fungal Disease. *PLoS Pathog.* **2023**, *19*, e1011322. [CrossRef] [PubMed]
84. Wu, S.; Song, R.; Liu, T.; Li, C. Antifungal Therapy: Novel Drug Delivery Strategies Driven by New Targets. *Adv. Drug Deliv. Rev.* **2023**, *199*, 114967. [CrossRef] [PubMed]
85. Maertens, J.; Marchetti, O.; Herbrecht, R.; Cornely, O.A.; Flückiger, U.; Frêre, P.; Gachot, B.; Heinz, W.J.; Lass-Flörl, C.; Ribaud, P.; et al. European guidelines for antifungal management in leukemia and hematopoietic stem cell transplant recipients: Summary of the ECIL 3—2009 Update. *Bone Marrow Transplant.* **2011**, *46*, 709–718. [CrossRef]
86. Cornely, O.A.; Bassetti, M.; Calandra, T.; Garbino, J.; Kullberg, B.J.; Lortholary, O.; Meersseman, W.; Akova, M.; Arendrup, M.C.; Arikan-Akdagli, S.; et al. ESCMID guideline for the diagnosis and management of Candida diseases 2012: Non-neutropenic adult patients. *Clin. Microbiol. Infect.* **2012**, *18* (Suppl. S7), 19–37. [CrossRef]

Journal of
Fungi

MDPI

Article

Patterns and Predictors of *Candida auris* Candidemia with Multidrug-Resistant Bacterial Co-Infections: Results from the CANDI-MDR Study

Karolina Akinosoglou [1,2,3,*,†], Katerina Skintzi [1,4,†], Ioannis Chandroulis [5], Eleni Polyzou [1,2], Argiro Siapika [6], Foteini Fligkou [1,7], Fotini Paliogianni [1,6], Charalambos Gogos [1] and George Dimopoulos [8]

1 Department of Medicine, University of Patras, Rio, 265 04 Patras, Greece; katrinski80@gmail.com (K.S.); polyzou.el@gmail.com (E.P.); fflig@yahoo.com (F.F.); fpal@med.upatras.gr (F.P.); cgogos@med.upatras.gr (C.G.)
2 Department of Internal Medicine, University General Hospital of Patras, Rio, 265 04 Patras, Greece
3 Division of Infectious Diseases, University General Hospital of Patras, Rio, 265 04 Patras, Greece
4 Department for Infection Control, University General Hospital of Patras, Rio, 265 04 Patras, Greece
5 School of Science and Technology, Hellenic Open University, 263 31 Patras, Greece; std528050@ac.eap.gr
6 Department of Microbiology, University General Hospital of Patras, Rio, 265 04 Patras, Greece; i.siapika@gmail.com
7 Department of Anesthesiology and Intensive Care Medicine, University General Hospital of Patras, Rio, 265 04 Patras, Greece
8 3rd Department of Critical Care, Evgenidio Hospital, Medical School, National and Kapodistrian University of Athens, 115 28 Athens, Greece; gdimop@med.uoa.gr
* Correspondence: akin@upatras.gr
† These authors equally contributed to this work.

Abstract: Introduction: *Candida auris* (now *Candidozyma auris*) and multidrug-resistant (MDR) bacterial infections pose significant therapeutic challenges due to high antimicrobial resistance, increased mortality, and persistence in healthcare settings. In Greece, their rising prevalence is raising concerns regarding co-infection, yet comprehensive data remain limited. This study aims to investigate the epidemiology, risk factors, and clinical outcomes of MDR bacterial co-infection in patients with *C. auris* candidemia. Methods: This single-center, retrospective observational cohort study was conducted at a Greek tertiary university hospital and included adult patients with *C. auris* bloodstream infections from January 2019 to June 2024. The data were analyzed using appropriate statistical methodologies. Results: Among 96 patients, those with *C. auris* candidemia and MDR bacterial co-infection exhibited a significantly higher mortality rate (87.23% vs. 61.22%, $p = 0.007$). The presence of a central venous catheter was the only factor significantly associated with MDR co-infection ($p = 0.030$). In univariate analysis, MDR co-infection, a higher Charlson Comorbidity Index, and mechanical ventilation correlated with increased mortality. Multivariate analysis identified MDR co-infection (OR = 3.19, $p = 0.045$) and mechanical ventilation (OR = 7.07, $p = 0.002$) as independent mortality predictors. Conclusions: These findings underscore the need for enhanced surveillance, precise identification, and stringent infection control measures to prevent *C. auris* and MDR bacterial outbreaks in healthcare settings.

Keywords: *Candida auris*; *Candidozyma auris*; bacteremia; candidemia; co-infection; MDR

1. Introduction

Candida auris (now *Candidozyma auris*) is an emerging multidrug-resistant (MDR) fungal pathogen that has become a global health threat since its first description in 2009 [1]. It has been reported in over 61 countries across all six inhabited continents [2,3]. *C. auris* affects both adults and children with various risk factors, including immunosuppression, diabetes, recent antibiotic use, catheter use, and prolonged hospital stays [2,4,5]. The pathogen is difficult to identify using traditional methods and can be misidentified as other yeasts [4]. The fungus exhibits high resistance to fluconazole and moderate resistance to amphotericin B and caspofungin [2]. *C. auris* infections are associated with a 30-day mortality rate of 39.5%, with bloodstream infections having a higher mortality of 45% [2]. The pathogen's ability to persist in hospital environments and resist decontamination procedures has led to outbreaks in healthcare facilities [6]. Hence, prompt investigation, aggressive interventions, and a robust response involving laboratories, clinicians, and public health agencies are crucial to manage cases and prevent transmission [7]

In this context, MDR bacterial infections pose a significant global health threat, with increasing prevalence and limited treatment options [8–10]. Antibiotic overuse in human therapies, animal husbandry, and aquaculture, as well as, poor antimicrobial stewardship and infection control practices [9], has contributed to rise of infections by MDR pathogens. The ESKAPE pathogens, including *Enterococcus faecium, Staphylococcus aureus, Klebsiella pneumoniae, Acinetobacter baumannii, Pseudomonas aeruginosa,* and *Enterobacter* spp., are particularly concerning, due to their role in life-threatening nosocomial infections [11]. Studies have identified various risk factors for MDR organisms, including diabetes mellitus, previous antibiotic use or hospitalization, use of medical devices, immunosuppressive therapy, and so on [12]. MDR infections are associated with increased morbidity, mortality, and healthcare costs [13]. Patients with MDR infections also had higher odds of 30-day readmission compared to those with non-MDR infections [14]. The rising prevalence of MDR infections presents challenges in selecting empiric antimicrobial therapy, necessitating a multidisciplinary approach involving antimicrobial stewardship and infection control practices to prevent further spread [9,10].

C. auris has emerged as a significant healthcare concern in Greece since its first isolation in 2019 [15]. The pathogen has rapidly spread across Greek healthcare facilities, with outbreaks reported in multiple hospitals [16,17]. *C. auris* isolates in Greece predominantly belong to the South Asian clade I and exhibit high resistance to fluconazole [16,18]. The fungus has become a leading cause of candidemia in some hospitals, with mortality rates exceeding 50% [16,17]. Of particular concern is the emergence of pan-echinocandin resistant isolates [17]. Similarly, Greece faces a significant challenge with MDR bacterial infections, particularly in hospital settings. A study of *P. aeruginosa* isolates from ventilator-associated pneumonia patients found high rates of MDR, extensively drug-resistant (XDR), and pandrug-resistant (PDR) strains, especially in Greece [19]. Recent surveillance data showed a rapid increase in MDR *P. aeruginosa* prevalence, rising by 63.8% from 2020 to 2023 [20]. MDR rates in Greece are among the highest in Europe, with carbapenem-resistant Gram-negative species becoming endemic in intensive care units [21]. A study of surgical ward patients found that 46.5% of isolated Gram-negative bacilli were MDR, with *A. baumannii* showing the highest MDR rate at 83.9% [22].

Of note, COVID-19 pandemic has led to profound shifts in healthcare delivery, with both immediate and lasting effects on individuals and communities. According to the CDC, contributing factors include a shift in patient demographics toward more severely ill individuals, reduced access to healthcare services leading to underdiagnosis and misdiagnosis, increased reliance on over-the-counter medications, and widespread antibiotic overuse. Additionally, challenges in laboratory supplies, diagnostic testing, and treatment availability have further complicated care. Crucially, the pandemic disrupted key infection prevention and control practices, such as hand hygiene, equipment disinfection, patient isolation, and consistent personal protective equipment use, ultimately undermining previous progress in addressing antimicrobial resistance (AMR). As a result, changes in hospitals' daily practice due to the COVID-19 pandemic had a severe impact on AMR [23]. A reduction in the detection and reporting of AMR data was documented during the pandemic, along with an alarming increase in resistant infections during hospitalization [24–27]

Both *C. auris* and MDR bacterial pathogens have several common predisposing factors for colonization and invasive infections. However, there are limited data on co-infection, as well as, on associated patterns and outcomes. A recent report has noted a high rate of co-colonization with other MDR organisms among *C. auris* carriers, particularly in long-term acute care hospitals [28]; however, no information is available on co-infections. The primary objective of this study was to investigate the epidemiology of co-infection with MDR pathogens in patients with *C. auris* candidemia and to identify associated risk factors. The secondary objective included describing the clinical characteristics, antifungal susceptibility patterns, and outcomes of co-infection with MDR bacteria and *C. auris* candidemia.

2. Methods

2.1. Study Design, Patient Population, and Data Collection

This was a single-center, retrospective observational cohort study conducted in a Greek tertiary university hospital. Successive hospitalized adult patients diagnosed with bloodstream infection (primary or secondary) due to *C. auris* from 1 January 2019 to 1 June 2024, were included. Patient data were anonymously and securely recorded from medical records, including demographic characteristics, clinical parameters, laboratory values, microbiological data and outcomes. Patients with missing data files or evidence of commensal pathogens, indicative of contamination, were excluded from this analysis.

2.2. Microbiology Identification and Definitions

The initial identification of *C. auris* was performed through culture on chromogenic agar, followed by confirmation using MALDI-TOF MS (Bruker Daltonics, Billerica, MA, USA). Antifungal susceptibility testing was performed using the Etest gradient diffusion method (bioMérieux, Craponne, France). In *vitro* susceptibility of the bloodstream isolates was further evaluated using the Clinical and Laboratory Standards Institute (CLSI) reference broth microdilution method. The interpretation of resistance was based on the tentative breakpoints provided by the United States Centers for Disease Control and Prevention (CDC) for amphotericin B, fluconazole, and echinocandins, while species-specific CLSI epidemiological cut-off values were applied for the remaining azole antifungals [29,30].

Bacterial pathogens were identified using VITEK® 2 identification cards (bioMérieux, Craponne, France). This study assessed susceptibility to various antimicrobial agents, as well as, the minimum inhibitory concentrations (MICs). Antimicrobials were chosen based on guidelines from the European Committee on Antimicrobial Susceptibility Testing (EUCAST) and recommendations from the European Centre for Disease Prevention and Control (ECDC) and the CDC. The MICs for colistin were determined using the broth microdilution method (SensiTest™ Colistin, Liofilchem, Italy), following EUCAST guidelines to ensure reliable results for this particular agent. Interpretation of the MIC values was conducted in accordance with the EUCAST standards applicable at the time of testing, categorizing isolates as susceptible, intermediate, or resistant. MDR was defined based on criteria from the ECDC and the CDC, as non-susceptible to at least one agent in three or more antimicrobial classes.

2.3. Management Protocols

Patients with suspected MDR bacterial infections were managed according to institutional protocols aligned with the Infectious Diseases Society of America (IDSA) and European Society of Clinical Microbiology and Infectious Diseases (ESCMID) guidelines [31,32]. Empiric antimicrobial therapy was initiated based on clinical presentation, the site of infection, history of prior antimicrobial therapy, tolerance or allergy, local resistance patterns, and infection severity. In cases of suspected or confirmed *C. auris*, initial antifungal therapy with echinocandins was employed [29]. Patients with confirmed or suspected MDR bacterial infections or *C. auris* were placed under strict contact precautions in single-patient rooms, including the use of gowns and gloves by all healthcare personnel upon room entry, with dedicated medical equipment where possible. Enhanced environmental cleaning protocols were implemented, especially for *C. auris*. All cases were evaluated by an infectious diseases specialist within 24 h of initiation of therapy. Upon availability of microbiological identification and antimicrobial susceptibility testing, targeted de-escalation was performed in accordance with antimicrobial stewardship principles to optimize treatment efficacy and minimize resistance development.

2.4. Statistical Methods

Categorical variables are presented as frequencies and relative frequencies, while continuous variables are presented as medians and interquartile ranges. Pearson's chi-squared test (χ^2) was used to assess associations between the categorical variables, with Yates's correction for continuity applied where appropriate. In cases where the data structure did not permit the use of Pearson's chi-squared test, Fisher's exact test was employed.

Shapiro–Wilk and Kolmogorov–Smirnov tests were used to assess the normality of continuous variables. Normally distributed continuous variables were compared using Student's t-test, whereas the Mann–Whitney U test was used for non-parametric comparisons.

In the logistic regression model for the multivariable evaluation of each independent variable's effect, the area under the receiver operating characteristic (ROC) curve (AUC) was calculated to assess the model's discriminative ability, with an AUC value greater than 0.7 considered acceptable. Additionally, pseudo R-squared statistics (McFadden's R^2, Cox & Snell R^2, and Nagelkerke R^2) were used to evaluate the model fit. Multicollinearity was assessed using the variance inflation factor (VIF), and a VIF < 5 was confirmed for all included variables. The selection of variables in the model was based on clinical relevance and statistical significance in univariable analysis.

All results are reported with 95% confidence intervals (95% CI). All tests were two-tailed, and statistical significance was set at $p = 0.05$.

Data analysis and visualization were performed in RStudio 2023.06.1, Build 524 (PBC, Boston, MA, USA), using the following R packages: "dplyr" v.1.1.4, "ggplot2" v.3.5.0, "scales" v.1.3.0, "nortest" v.1.0.4, "epiDisplay" v.3.5.0.2, "broom" v.1.0.5, "lmtest" v.0.9-40, "car" v.3.1-2, "stats" v.4.3.1, "ResourceSelection" v.0.3-6, "pROC" v.1.18.5, and "pscl" v.1.5.9.

2.5. Ethical Approval

The study was conducted according to the Declaration of Helsinki and Good Clinical Research Practice guidelines, and received approval from the local Ethics Committee and Institutional Review Board (391/31 July 2024). Informed consent was waived due to the retrospective nature of the study and the anonymous recording of data according to general data protection regulations.

3. Results

3.1. General Characteristics and Medical History of Participants

A total of 96 patients with candidemia (*C. auris*) were included in this study; their characteristics are shown in Table 1. Among the 96 hospitalized patients with candidemia, 48.96% (47 patients) also had a co-infection (at least one episode of bacteremia) with MDR bacterial pathogens. No statistically significant differences were observed between the two groups in terms of general characteristics, diagnosis on admission, or risk factors for MDR or *C. auris* infection/colonization. The only statistically significant difference was noted in the presence of a central venous catheter (CVC), which was more common among patients with co-infection by MDR pathogens (p = 0.030). The remaining risk factors did not show significant differences between the two groups.

Table 1. General characteristics and medical history of participants.

	Candidemia		
	Without Co-Infection (n = 49)	**With MDR Co-Infection (n = 47)**	***p***
GENERAL CHARACTERISTICS			
Age (Years)	69 (57–76)	70 (64–77)	0.392
Gender (Male)	26 (53.06 %)	30 (63.83 %)	0.388
CCI	4 (3–5)	4 (3–6)	0.372
Sepsis	12 (24.49%)	12 (25.53 %)	1
DIAGNOSIS			
COVID-19	8 (16.33%)	10 (21.28%)	0.719
RTI	18(36.73%)	17(36.17%)	1
UTI	4(8.16%)	3(6.38%)	1
Cerebrovascular Events	9(18.37%)	8(17.02%)	1
Cardiovascular Events	1(2.04%)	1(2.13%)	1
Trauma	8(16.33%)	6(12.77%)	0.837
Other	1(2.04%)	2(4.26%)	0.481
RISK FACTORS for MDR/*C. auris* Infection			
CVC	32 (66.67%)	41 (87.23%)	0.030
Foley Catheter	43 (87.76%)	44 (93.62%)	0.487
PICC	22 (44.9 %)	17 (36.17%)	0.507
IMV	24 (48.98%)	29 (61.7%)	0.294
Non-IMV	1 (2.04%)	2 (4.26%)	0.613
History of Prior Antifungal Treatment	28 (57.14%)	21 (44.68%)	0.292
History of Prior Antibiotic Treatment	47 (95.92%)	45 (95.74%)	1
Recent Hospitalization, within 6 Months (Yes)	34 (69.39%)	30 (63.83 %)	0.718
Pre-Existing *Candida* Colonization	21 (42.86%)	19 (40.43%)	0.972

CCI: Charlson Comorbidity Index; RTI: respiratory tract infections; UTI: urinary tract infections; MDR: multidrug-resistant pathogen; CVC: central venous catheter; PICC: peripherally inserted central catheter; IMV: invasive mechanical ventilation.

As for *C. auris*, no difference in the resistance patterns or antifungal regimens used were noted between the groups (Supplementary Table S1). As expected, the majority of isolates were resistant to fluconazole (approximately 98%), followed by amphotericin B

(24%), caspofungin (10%), anidulafungin (9%), and voriconazole (8%). More than 87% of cases were treated with anidulafungin or micafungin, while approximately 4% required therapy with amphotericin B, either alone or in combination with echinocandin.

Among the participants with co-infection by MDR pathogens, 89.36% of the infections were monomicrobial, whereas 10.64% were polymicrobial (Supplementary Table S2). From the total number of isolated pathogens (n = 52), 34.04% were Enterococci, of which 93.75% were vancomycin-resistant (VRE). Other isolated pathogens included *Acinetobacter* spp. (23.4%), *Klebsiella pneumoniae* (19.15%), *Pseudomonas* spp. (12.77%), *E. coli* (4.26%), other *Candida* spp. (12.77%), and miscellaneous pathogens (4.26%). Regarding antimicrobial resistance, carbapenemase-producing *K. pneumoniae* (KPC) was detected in 7.69% of cases. Additionally, metallo-β-lactamases (NDM, VIM) were identified in approximately 15% of the isolated strains. In contrast, extended-spectrum β-lactamases (ESBLs) were not detected (Supplementary Table S2).

3.2. Laboratory Findings

The laboratory parameters did not exhibit any statistically significant differences between the participant groups (Supplementary Table S3).

3.3. Outcomes

Regarding patient outcomes, the participants with concomitant co-infection with MDR pathogens exhibited a statistically significantly higher mortality rate (87.23% vs. 61.22%, $p = 0.007$). Conversely, post-infection colonization was marginally more frequent in the group with isolated candidemia (22.45% vs. 6.38%, $p = 0.05$). No other significant differences in outcome characteristics were observed between the two groups (Table 2).

Table 2. Outcomes of participants.

	Candidemia		
	Without Co-Infection (n = 49)	**With MDR Co-Infection (n = 47)**	*p*
Intensive Care Unit Admission (Yes)	23 (46.94%)	25 (53.19%)	0.683
Mortality	30 (61.22%)	41 (87.23%)	0.007
28-Day Mortality	19 (63.33%)	28 (68.29%)	0.066
90-Day Mortality	11 (36.67%)	13 (31.71%)	0.723
Death Associated with *C. auris*	27 (90%)	34 (82.93%)	0.501
Candidemia Resolution	38 (77.55%)	30 (63.83%)	0.209
Time to Bloodstream *Candida* Clearance (Days)	5 (2–9.75)	6 (3.25–10)	0.464
C. auris Relapse	16 (32.65%)	11 (23.4%)	0.435
Post-Infection *Candida* Colonization	11 (22.45%)	3 (6.38%)	0.050
LOS (Days)	68 (30–130)	46 (32.5–105.5)	0.465

MDR: Multidrug-resistant pathogens; LOS: length of stay.

3.4. Candidemia with MDR Co-Infection as an Independent Mortality Risk Factor

The potential associations between mortality outcomes and candidemia with MDR co-infection, age, length of hospital stay (LOS), CCI, and invasive mechanical ventilation (IMV) were analyzed. In the univariate analysis, candidemia with MDR co-infection was significantly associated with an increased risk of mortality (OR = 4.33, 95% CI: 1.54–12.14, *p*

= 0.005). Additionally, a higher CCI score (OR = 1.29, 95% CI: 1.01–1.65, *p* = 0.038) and IMV (OR = 4.73, 95% CI: 1.74–12.86, *p* = 0.002) were also significantly correlated with mortality. Age exhibited a borderline but non-significant association (OR = 1.03, 95% CI: 1–1.06, *p* = 0.062), while LOS was not statistically significant (*p* = 0.529). In the multivariate analysis, candidemia with MDR co-infection remained an independent predictor of mortality (OR = 3.19, 95% CI: 1.03–9.9, *p* = 0.045), along with IMV, which showed a strong independent association (OR = 7.07, 95% CI: 2.03–24.67, *p* = 0.002). Conversely, age (*p* = 0.864), LOS (*p* = 0.809), and CCI (*p* = 0.137) did not retain statistical significance after adjustment. These findings, summarized in Table 3, provide insight into both univariate and multivariate analyses, highlighting the significant role of MDR co-infection and intubation in patient mortality.

Table 3. Univariate and multivariate analysis of factors associated with mortality risk.

	Unadjusted Analysis			**Adjusted Analysis**		
	Odds Ratio	**95% CI**	***p*-Value**	**Odds Ratio**	**95% CI**	***p*-Value**
Candidemia Status (with MDRs)	4.33	1.54–12.14	**0.005**	3.19	1.03–9.9	0.045
Age	1.03	1–1.06	0.062	1.005	0.95–1.07	0.864
LOS	1.00	1–1.01	0.529	1	0.99–1.01	0.809
CCI	1.29	1.01–1.65	**0.038**	1.43	0.89–2.29	0.137
IMV (Yes)	4.73	1.74–12.86	**0.002**	7.07	2.03–24.67	0.002

MDR: Multidrug-resistant pathogens; LOS: length of stay; CCI: Charlson Comorbidity Index; IMV: invasive mechanical ventilation.

4. Discussion

This study aimed to investigate the patterns of MDR pathogen bacteremia in patients with *C. auris* candidemia, identify associated risk factors, and assess their impact on clinical outcomes. Patients with *C. auris* candidemia, with and without MDR co-infection, exhibited comparable characteristics, except for the presence of CVCs, which was more prevalent in those with MDR co-infections. Mortality rates were significantly higher in patients with MDR co-infections, even though *C. auris* colonization occurred less frequently after the initial infection. IMV and co-infection with MDR pathogens were identified as significant predictors of increased mortality in patients with *C. auris* candidemia.

Our study is one of the largest cohorts of *C. auris* patients described in the global literature. Studies have reported outbreaks in Oman [33], the UK [34], and the USA [35], highlighting its global spread. The COVID-19 pandemic has led to an increase in *C. auris* infections, particularly among critically ill patients in ICUs [36,37]. *C. auris* has emerged as a significant threat in healthcare settings, with studies reporting it as the most isolated *Candida* species in blood cultures of COVID-19 patients [37]. Prior data indicate an increasing trend in *C. auris* colonization and clinical infections, particularly within long-term care settings, such as long-term acute care hospitals and ventilator-equipped skilled nursing facilities [38,39]. Key risk factors for *C. auris* colonization and candidemia include IMV, CVC, total parenteral nutrition, and prolonged ICU stays [40–42]. Prior antibiotic use, especially carbapenems, and antifungal exposure, particularly fluconazole, are also associated with increased risk [39,42]. Almost all our patients had priorly received antibiotics, whereas the majority had been recently hospitalized or were under invasive mechanical ventilation in our cohort.

Nearly half of the patients with *C. auris* candidemia also had MDR bacteremia. Concurrent fungemia and bacteremia represents a serious condition associated with critically ill patients, particularly in surgical settings. Studies have shown that polymicrobial fungemia accounts for 3.4% of fungemia cases, with *Candida* species being the most common fungal pathogens [43,44]. Predisposing factors include intravascular catheters, antibiotic use, and surgical procedures [44], while mixed septicemia, characterized by synchronous bacterial and fungal bloodstream infections, is a marker for critically ill surgical patients with high mortality rates [45]. The 50% incidence of MDR pathogens observed in this study is consistent with the high prevalence of MDR infections in Greece, as reported in previous studies, [21,46] and a recent report by the European Centre for Disease Prevention and Control (ECDC) [47]. Since the late 2000s, Greece has faced an endemic challenge of MDR pathogens within its hospital sector, predominantly driven by carbapenem-resistant Gram-negative bacilli; the country consistently reports some of the highest antimicrobial resistance rates in Europe [47].

Patients with *C. auris* infection, with or without MDR co-infection, did not differ in terms of risk factors. This is in line with previous data showing common risk factors for *C. auris* infection or colonization and MDR bacteremia, including the prior use of antibiotics, immunocompromise, indwelling catheters, IMV, and prior recent hospitalization [3,12,39]. However, patients with MDR co-infections had a higher prevalence of CVC use (87.23 vs. 66.67%). The presence of a CVC is associated with higher rates of bloodstream infections [48,49]. Central line-associated bloodstream infections (CLABSIs) are a significant healthcare concern, with MDR organisms causing up to 67% of cases [50]. Gram-negative bacteria are the predominant pathogens in some settings, including Greece, Taiwan, Malaysia, Saudi Arabia, and China, with 9% being MDR pathogens [51–55]. CLABSIs are associated with increased mortality, costs, and readmission rates. Failure to promptly remove CVC in cases of CLABSI caused by MDR pathogens has been strongly associated with increased 30-day mortality, as persistent infection and inadequate source control can significantly worsen patient outcomes [56]. Therefore, the implementation of standardized central line care bundles, including the proper insertion techniques, regular line maintenance, and timely removal when no longer necessary, remains a critical strategy to minimize the risk of CLABSI, particularly in high-risk populations vulnerable to MDR infections [51].

In our report, *C. auris* colonization was observed in approximately 40% and 15% of patients, prior to and following candidemia, respectively. Research indicates that *C. auris* colonization is prevalent in healthcare settings, especially among critically ill patients and residents of long-term care facilities [39,57,58]. *C. auris* can colonize multiple body sites, with the axilla and groin being common locations [57,59]. The persistence of *C. auris* on skin and environmental surfaces contributes to its transmission in healthcare settings [59]. Outbreaks have been associated with contaminated medical equipment, such as reusable temperature probes [60]. Effective control measures involve comprehensive patient and environmental screening, chlorhexidine washes for patients, and thorough surface disinfection with suitable agents, such as stabilized hydrogen peroxide [61]. When followed correctly, proper hand hygiene and commonly used disinfectants can be effective against *C. auris* [61]. Sodium hypochlorite isotonic solution has shown potential for patient decolonization [62]. However, *C. auris* can survive on dry fabrics for up to seven days, posing challenges for eradication [61]. Colonization with *C. auris* significantly increases the risk of subsequent invasive infection, often by the same strain [63]. We observed reduced post-infection colonization in patients with MDR co-infection. It is possible that, enhanced disinfectant and de-colonization measures associated with MDR bacteremia may have contributed to the (marginally) significant difference between the groups. However, this finding should be interpreted with caution. The CDC no longer recommends routine reassessments for *C. auris* colonization, partly due to the fluctuating nature of its detection, as *C. auris* can sometimes be present and at other times undetected, especially in healthcare settings [38]. To manage *C. auris* in healthcare settings, rigorous colonization screening and strict infection control measures are essential [64].

C. auris exhibits high resistance to fluconazole and moderate resistance to amphotericin B and caspofungin, while remaining sensitive to echinocandins, like micafungin and anidulafungin [2,65]. This resistance is attributed to multiple molecular mechanisms. Mutations in the ERG11 gene, increased copy numbers of this gene, and the overexpression of efflux pumps contribute to fluconazole resistance [66]. A novel genetic determinant, mutations in the TAC1B gene, has been identified as a significant contributor to clinical fluconazole resistance [67]. These mutations can arise rapidly upon exposure to fluconazole and are present in the majority of fluconazole-resistant *C. auris* isolates globally. Additionally, resistance to other antifungals has been observed, including amphotericin B (10–30% of strains) and echinocandins (up to 5% of strains) [68]. We also recorded high resistance to fluconazole in our study, in line with the fact that the majority of patients had been under antifungal treatment in the past.

The majority of patients are empirically treated with echinocandins, in accordance with international guidelines and local protocols [29]. This regimen is later modified if required, following antifungal sensitivity testing and patient individual needs, e.g., tolerance, allergies, drug–drug interactions etc. At the moment, *C. auris* commonly shows resistance to fluconazole and amphotericin B, with echinocandin resistance emerging in some regions [69]. In this context, several new antifungals are being developed to combat *C. auris*. APX001A, which inhibits the fungal protein Gwt1, demonstrated potent in vitro activity and improved survival in mouse models compared to anidulafungin [70]. Ibrexafungerp, a novel oral triterpenoid antifungal, showed broad *in vitro* activity against *C. auris*, including isolates with fks mutations, and promising clinical results in two patients [71]. Other investigational agents include rezafungin, manogepix/fosmanogepix, olorofim, and tetrazoles, which have shown improvements in survival and reduction in tissue fungal burden in animal models [72].

The overall mortality rate for *C. auris* infections ranges from 30% to 60%, with bloodstream infections having a 45% mortality rate [2]. However, a recent study suggests that mortality may be lower when adjusted for confounding variables [73]. In our cohort, mortality was approximately 60% in patients with *C. auris* candidemia. We found significantly increased mortality in *C. auris* patients with MDR co-infections compared to those without (87%), while MDR infection was identified as an independent risk factor for mortality. This finding is consistent with previous research, including ours, demonstrating increased mortality rates among patients with MDR infections [74–76]. However, it has been suggested that the elevated mortality risk associated with MDR pathogens is primarily attributed to the delayed initiation of appropriate antimicrobial therapy, rather than the intrinsic virulence of the resistant microorganisms themselves [74,77–79]. Even though the treatment protocols at our site align well with international guidelines on the management of MDR pathogens (see methods), no information is available on the details of the regimens and the timing used.

Invasive ventilation was identified as an independent predictor for mortality in these patients. Invasive mechanical ventilation in critically ill patients is strongly associated with high mortality rates, particularly in pediatric hematopoietic stem cell transplant recipients, with reported pediatric ICU mortality rates as high as 60.4% [80]. In patients with COVID-19-related acute respiratory distress syndrome (ARDS), factors such as elevated driving pressure and ventilatory rate have been identified as key contributors to increased mortality [81]. For patients without ARDS, maximum airway pressure has emerged as the primary respiratory factor, independently linked to in-hospital mortality [82]. Furthermore, early extubating to non invasive ventilation as part of a weaning strategy has been shown to reduce hospital mortality, particularly in patients with chronic obstructive pulmonary disease. This strategy contributes to a reduction in the duration of invasive ventilation and ICU length of stay, associated with a lower incidence of ventilator-associated pneumonia [83].

Despite being one of the largest *C. auris* cohorts described in the literature, our study suffers inherent limitations. First it was a retrospective single-center study; hence, our findings and conclusions drawn should be interpreted with caution. Diverse practices and increased local MDR endemicity limits generalization of our results. Moreover, misidentification of *C. auris* is common due to inadequate reference databases on diagnostic platforms [84]. Additionally, the correlation between MIC values and clinical outcomes is poorly understood, resulting in a lack of *C. auris*-specific breakpoints [69]. Although various clinical outcomes, such as mortality, LOS, and ICU admission, were assessed, this study lacked crucial information on several key variables that could significantly influence these endpoints. Specifically, no detailed data were available regarding the antibacterial treatment regimens employed, including type, dosage, duration, and timing of administration. Additionally, the absence of information on source control measures, such as surgical interventions, drainage procedures, and institutional disinfection practices, limits our ability to interpret the results fully. These factors are known to play a critical role in patient outcomes and may confound the relationship between the observed variables and the reported clinical outcomes.

Our study was the first to assess *C. auris* and bacterial MDR co-infection. No specific factors were identified in the *C. auris* patient group also presenting with MDR pathogens, except for the presence of CVC. MDR co-infections were associated with increased mortality, especially among patients under IMV. Our findings underscore the critical need for enhanced infection control measures in healthcare settings. Close monitoring and stringent catheter care protocols are essential, particularly among high-risk patients, while targeted antimicrobial stewardship, early diagnostic screening, and rigorous hygiene practices should be prioritized. Healthcare personnel must be regularly trained in recognizing and managing co-infections, while infection prevention teams should consider implementing routine surveillance for MDR pathogens in patients with *C. auris*. This can be challenging in low- or middle-income countries, while global mobility adds to the complexity, facilitating the cross-border spread of *C. auris* and MDR organisms. Global health bodies and national public health authorities must invest in building laboratory capacity, enforcing standardized infection prevention and control guidelines, and supporting antimicrobial stewardship programs across all healthcare levels. International collaboration, real-time data sharing, and coordinated surveillance efforts are essential to monitor trends, detect outbreaks early, and mount effective, evidence-based responses to this dual threat.

5. Conclusions

Our findings underscore the urgent need for enhanced surveillance, accurate identification methods, and stringent infection control measures to combat *C. auris*, prevent new infections with MDR pathogens and avoid epidemic outbreaks.

Supplementary Materials: The following supporting information can be downloaded at: https://www.mdpi.com/article/10.3390/jof11060407/s1, Table S1. *C. auris* resistance patterns and employed therapy. Table S2. Microbiological profile and antimicrobial resistance in patients with candidemia and MDR co-infection. Table S3. Laboratory parameters.

Author Contributions: K.A. and C.G. conceived the idea; K.S. and E.P. collected the patient data; A.S. and F.P. collected the microbiology data; I.C. and K.A. analyzed the data; K.A. and I.C. wrote the manuscript and drew the figures and tables; F.F., F.P., C.G., and G.D. corrected the manuscript; K.A. oversaw the study; K.A. revised the manuscript. All authors have read and agreed to the published version of the manuscript.

Funding: This research received no external funding.

Institutional Review Board Statement: The study was conducted in accordance with the Declaration of Helsinki, and approved by the Institutional Review Board and Ethics Committee of University General Hospital of Patras (391/31 July 2024).

Informed Consent Statement: Patient consent was waived due to retrospective nature of the study and the anonymous recording of data according to general data protection regulations.

Data Availability Statement: The original contributions presented in this study are included in the article/Supplementary Materials. Further inquiries can be directed to the corresponding author.

Conflicts of Interest: The authors declare no conflicts of interest.

Abbreviations

MDR	Multidrug-resistant
PDR	Pandrug-resistant
AMR	Antimicrobial resistance
CVC	Central venous catheter
RTI	Respiratory tract infections
UTI	Urinary tract infections
VRE	Vancomycin-resistant Enterococci
ESBLs	Extended-spectrum β lactamases
KPC	*K. pneumoniae* carbapenemase
MBL	Metallo-β-lactamases
NDM	New Delhi metallo-β-lactamase
VIM	Verona integron-encoded metallo-β-lactamase
LOS	Length of stay
CCI	Charlson Comorbidity Index
IMV	Invasive mechanical ventilation
ECDC	European Centre for Disease Prevention and Control
CLABSIs	Central line-associated bloodstream infections
ICU	Intensive care unit
ARDS	Acute respiratory distress syndrome

References

1. Du, H.; Bing, J.; Hu, T.; Ennis, C.L.; Nobile, C.J.; Huang, G. *Candida auris*: Epidemiology, biology, antifungal resistance, and virulence. *PLoS Pathog.* **2020**, *16*, e1008921. [CrossRef] [PubMed]
2. Chen, J.; Tian, S.; Han, X.; Chu, Y.; Wang, Q.; Zhou, B.; Shang, H. Is the superbug fungus really so scary? A systematic review and meta-analysis of global epidemiology and mortality of *Candida auris*. *BMC Infect Dis.* **2020**, *20*, 827. [CrossRef] [PubMed]
3. Bhargava, A.; Klamer, K.; Sharma, M.; Ortiz, D.; Saravolatz, L. *Candida auris*: A Continuing Threat. *Microorganisms* **2025**, *13*, 652. [CrossRef]
4. Jeffery-Smith, A.; Taori, S.K.; Schelenz, S.; Jeffery, K.; Johnson, E.M.; Borman, A.; *Candida auris* Incident Management Team; Manuel, R.; Brown, C.S. Candida auris: A Review of the Literature. *Clin. Microbiol. Rev.* **2018**, *31*, e00029-17. [CrossRef]
5. Sanyaolu, A.; Okorie, C.; Marinkovic, A.; Abbasi, A.F.; Prakash, S.; Mangat, J.; Hosein, Z.; Haider, N.; Chan, J. *Candida auris*: An Overview of the Emerging Drug-Resistant Fungal Infection. *Infect. Chemother.* **2022**, *54*, 236–246. [CrossRef]
6. Ahmad, S.; Alfouzan, W. *Candida auris*: Epidemiology, Diagnosis, Pathogenesis, Antifungal Susceptibility, and Infection Control Measures to Combat the Spread of Infections in Healthcare Facilities. *Microorganisms* **2021**, *9*, 807. [CrossRef]

7. Forsberg, K.; Woodworth, K.; Walters, M.; Berkow, E.L.; Jackson, B.; Chiller, T.; Vallabhaneni, S. *Candida auris*: The recent emergence of a multidrug-resistant fungal pathogen. *Med. Mycol.* **2019**, *57*, 1–12. [CrossRef] [PubMed]
8. Theuretzbacher, U. Global antibacterial resistance: The never-ending story. *J. Glob. Antimicrob. Resist.* **2013**, *1*, 63–69. [CrossRef]
9. Cerceo, E.; Deitelzweig, S.B.; Sherman, B.M.; Amin, A.N. Multidrug-Resistant Gram-Negative Bacterial Infections in the Hospital Setting: Overview, Implications for Clinical Practice, and Emerging Treatment Options. *Microb. Drug Resist.* **2016**, *22*, 412–431. [CrossRef]
10. Collaborators GBDAR. Global burden of bacterial antimicrobial resistance 1990–2021: A systematic analysis with forecasts to 2050. *Lancet* **2024**, *404*, 1199–1226. [CrossRef]
11. Esposito, S.; De Simone, G. Update on the main MDR pathogens: Prevalence and treatment options. *Infez. Med.* **2017**, *25*, 301–310. [PubMed]
12. Patolia, S.; Abate, G.; Patel, N.; Patolia, S.; Frey, S. Risk factors and outcomes for multidrug-resistant Gram-negative bacilli bacteremia. *Ther. Adv. Infect. Dis.* **2018**, *5*, 11–18. [CrossRef] [PubMed]
13. van Duin, D.; Paterson, D.L. Multidrug-Resistant Bacteria in the Community: Trends and Lessons Learned. *Infect. Dis. Clin. North. Am.* **2016**, *30*, 377–390. [CrossRef]
14. Dicks, K.V.; Anderson, D.J.; Baker, A.W.; Sexton, D.J.; Lewis, S.S. Clinical Outcomes and Healthcare Utilization Related to Multidrug-Resistant Gram-Negative Infections in Community Hospitals. *Infect. Control Hosp. Epidemiol.* **2017**, *38*, 31–38. [CrossRef]
15. Stathi, A.; Loukou, I.; Kirikou, H.; Petrocheilou, A.; Moustaki, M.; Velegraki, A.; Zachariadou, L. Isolation of *Candida auris* from cystic fibrosis patient, Greece, April 2019. *Euro Surveill.* **2019**, *24*, 7–11. [CrossRef] [PubMed]
16. Politi, L.; Vrioni, G.; Hatzianastasiou, S.; Lada, M.; Martsoukou, M.; Sipsas, N.V.; Chini, M.; Baka, V.; Kafkoula, E.; Masgala, A.; et al. *Candida auris* in Greek healthcare facilities: Active surveillance results on first cases and outbreaks from eleven hospitals within Attica region. *J. Mycol. Med.* **2024**, *34*, 101477. [CrossRef]
17. Meletiadis, J.; Siopi, M.; Spruijtenburg, B.; Georgiou, P.-C.; Kostoula, M.; Vourli, S.; Frantzeskaki, F.; Paramythiotou, E.; Meis, J.F.; Tsangaris, I.; et al. *Candida auris* fungaemia outbreak in a tertiary care academic hospital and emergence of a pan-echinocandin resistant isolate, Greece, 2021 to 2023. *Euro Surveill.* **2024**, *29*, 2400128. [CrossRef]
18. Poulopoulou, A.; Sidiropoulou, A.; Sarmourli, T.; Zachrou, E.; Michailidou, C.; Zarras, C.; Vagdatli, E.; Massa, E.; Mouloudi, E.; Pyrpasopoulou, A.; et al. *Candida auris*: Outbreak, surveillance and epidemiological monitoring in Northern Greece. *Med. Mycol.* **2024**, *62*, myae062. [CrossRef]
19. Pérez, A.; Gato, E.; Pérez-Llarena, J.; Fernández-Cuenca, F.; Gude, M.J.; Oviaño, M.; Pachón, M.E.; Garnacho, J.; González, V.; Pascual, Á.; et al. High incidence of MDR and XDR *Pseudomonas aeruginosa* isolates obtained from patients with ventilator-associated pneumonia in Greece, Italy and Spain as part of the MagicBullet clinical trial. *J. Antimicrob. Chemother.* **2019**, *74*, 1244–1252. [CrossRef]
20. Polemis, M.; Theofili, E.; Tryfinopoulou, K. Rapid increase of multi-drug resistant *Pseudomonas aeruginosa* in Greece—WHONET-Greece (2020–2023). *Eur. J. Public Health* **2024**, *34*, ckae144.2154. [CrossRef]
21. Miyakis, S.; Pefanis, A.; Tsakris, A. The challenges of antimicrobial drug resistance in Greece. *Clin. Infect. Dis.* **2011**, *53*, 177–184. [CrossRef] [PubMed]
22. Maraki, S.; Mavros, M.N.; Kofteridis, D.P.; Samonis, G.; Falagas, M.E. Epidemiology and antimicrobial sensitivities of 536 multi-drug-resistant gram-negative bacilli isolated from patients treated on surgical wards. *Surg. Infect.* **2012**, *13*, 326–331. [CrossRef]
23. Polemis, M.; Mandilara, G.; Pappa, O.; Argyropoulou, A.; Perivolioti, E.; Koudoumnakis, N.; Pournaras, S.; Vasilakopoulou, A.; Vourli, S.; Katsifa, H.; et al. COVID-19 and Antimicrobial Resistance: Data from the Greek Electronic System for the Surveillance of Antimicrobial Resistance-WHONET-Greece (January 2018–March 2021). *Life* **2021**, *11*, 996. [CrossRef]
24. Micheli, G.; Sangiorgi, F.; Catania, F.; Chiuchiarelli, M.; Frondizi, F.; Taddei, E.; Murri, R. The Hidden Cost of COVID-19: Focus on Antimicrobial Resistance in Bloodstream Infections. *Microorganisms* **2023**, *11*, 1299. [CrossRef]
25. Polly, M.; de Almeida, B.L.; Lennon, R.P.; Cortes, M.F.; Costa, S.F.; Guimaraes, T. Impact of the COVID-19 pandemic on the incidence of multidrug-resistant bacterial infections in an acute care hospital in Brazil. *Am. J. Infect. Control.* **2022**, *50*, 238–239. [CrossRef]
26. Yang, X.; Li, X.; Qiu, S.; Liu, C.; Chen, S.; Xia, H.; Zeng, Y.; Shi, L.; Chen, J.; Zheng, J.; et al. Global antimicrobial resistance and antibiotic use in COVID-19 patients within health facilities: A systematic review and meta-analysis of aggregated participant data. *J. Infect.* **2024**, *89*, 106183. [CrossRef]
27. Attal, R.; Deotale, V. Bacterial co-infections and secondary infections and their antimicrobial resistance in Covid-19 patients during the second pandemic wave. *GMS Hyg. Infect. Control* **2024**, *19*, Doc10. [PubMed]

28. Barza, R.; Patel, P.; Patel, J.; [illegible]; [illegible]; Mangold, K.; Lepinski, J.; Hines, D.; Lee, B.; Singh, K. 1687. High-Rates of *Candida auris* Carriage and Co-Colonization with Multidrug-Resistant Organisms (MDROs). *Open Forum Infect. Dis.* [illegible], 23, S617. [CrossRef]
29. Centres for Disease Control and Prevention. *Candida auris*. 2024. Available online: https://www.cdc.gov/candida-auris/hcp/laboratories/antifungal-susceptibility-testing.html (accessed on 13 March 2025).
30. Arendrup, M.C.; Prakash, A.; Meletiadis, J.; Sharma, C.; Chowdhary, A. Comparison of EUCAST and CLSI Reference Microdilution MICs of Eight Antifungal Compounds for *Candida auris* and Associated Tentative Epidemiological Cutoff Values. *Antimicrob. Agents Chemother.* **2017**, *61*, e00485-17. [CrossRef]
31. Tamma, P.D.; Heil, E.L.; Justo, J.A.; Mathers, A.J.; Satlin, M.J.; Bonomo, R.A. Infectious Diseases Society of America 2024 Guidance on the Treatment of Antimicrobial-Resistant Gram-Negative Infections. *Clin. Infect. Dis.* **2024**, ciae403. [CrossRef]
32. Paul, M.; Carrara, E.; Retamar, P.; Tängdén, T.; Bitterman, R.; Bonomo, R.A.; de Waele, J.; Daikos, G.L.; Akova, M.; Harbarth, S.; et al. European Society of Clinical Microbiology and Infectious Diseases (ESCMID) guidelines for the treatment of infections caused by multidrug-resistant Gram-negative bacilli (endorsed by European society of intensive care medicine). *Clin. Microbiol. Infect.* **2022**, *28*, 521–547. [CrossRef] [PubMed]
33. Al Maani, A.; Paul, H.; Al-Rashdi, A.; Wahaibi, A.A.; Al-Jardani, A.; Al Abri, A.M.A.; AlBalushi, M.A.H.; Al-Abri, S.; Al Reesi, M.; Al Maqbali, A.; et al. Ongoing Challenges with Healthcare-Associated *Candida auris* Outbreaks in Oman. *J. Fungi* **2019**, *5*, 101. [CrossRef] [PubMed]
34. Schelenz, S.; Hagen, F.; Rhodes, J.L.; Abdolrasouli, A.; Chowdhary, A.; Hall, A.; Ryan, L.; Shackleton, J.; Trimlett, R.; Meis, J.F.; et al. First hospital outbreak of the globally emerging *Candida auris* in a European hospital. *Antimicrob. Resist. Infect. Control.* **2016**, *5*, 35. [CrossRef]
35. Adams, E.; Quinn, M.; Tsay, S.; Poirot, E.; Chaturvedi, S.; Southwick, K.; Greenko, J.; Fernandez, R.; Kallen, A.; Vallabhaneni, S.; et al. *Candida auris* in Healthcare Facilities, New York, USA, 2013-2017. *Emerg. Infect. Dis.* **2018**, *24*, 1816–1824. [CrossRef] [PubMed]
36. Rahman, M.A.; Victoros, E.; Shanjana, Y.; Thomas, M.R.; Islam, M.R. The *Candida auris* Infection After the COVID-19 Pandemic Seems to be an Urgent Public Health Emergency: A Call to Attention. *Health Sci. Rep.* **2024**, *7*, e70160. [CrossRef]
37. Vinayagamoorthy, K.; Pentapati, K.C.; Prakash, H. Prevalence, risk factors, treatment and outcome of multidrug resistance *Candida auris* infections in Coronavirus disease (COVID-19) patients: A systematic review. *Mycoses* **2022**, *65*, 613–624. [CrossRef]
38. Pacilli, M.; Kerins, J.L.; Clegg, W.J.; Walblay, K.A.; Adil, H.; Kemble, S.K.; Xydis, S.; McPherson, T.D.; Lin, M.Y.; Hayden, M.K.; et al. Regional Emergence of *Candida auris* in Chicago and Lessons Learned From Intensive Follow-up at 1 Ventilator-Capable Skilled Nursing Facility. *Clin. Infect. Dis.* **2020**, *71*, e718–e725. [CrossRef]
39. Rossow, J.; Ostrowsky, B.; Adams, E.; Greenko, J.; McDonald, R.; Vallabhaneni, S.; Forsberg, K.; Perez, S.; Lucas, T.; Alroy, K.A.; et al. Factors Associated with *Candida auris* Colonization and Transmission in Skilled Nursing Facilities with Ventilator Units, New York, 2016–2018. *Clin. Infect. Dis.* **2021**, *72*, e753–e760. [CrossRef]
40. Ruiz-Gaitan, A.; Martinez, H.; Moret, A.M.; Calabuig, E.; Tasias, M.; Alastruey-Izquierdo, A.; Zaragoza, Ó.; Mollar, J.; Frasquet, J.; Salavert-Lletí, M.; et al. Detection and treatment of *Candida auris* in an outbreak situation: Risk factors for developing colonization and candidemia by this new species in critically ill patients. *Expert. Rev. Anti Infect. Ther.* **2019**, *17*, 295–305. [CrossRef]
41. Garcia-Bustos, V.; Salavert, M.; Ruiz-Gaitan, A.C.; Cabanero-Navalon, M.D.; Sigona-Giangreco, I.A.; Peman, J. A clinical predictive model of candidaemia by *Candida auris* in previously colonized critically ill patients. *Clin. Microbiol. Infect.* **2020**, *26*, 1507–1513. [CrossRef]
42. Shastri, P.S.; Shankarnarayan, S.A.; Oberoi, J.; Rudramurthy, S.M.; Wattal, C.; Chakrabarti, A. *Candida auris* candidaemia in an intensive care unit—Prospective observational study to evaluate epidemiology, risk factors, and outcome. *J. Crit. Care* **2020**, *57*, 42–48. [CrossRef] [PubMed]
43. Guerra-Romero, L.; Telenti, A.; Thompson, R.L.; Roberts, G.D. Polymicrobial fungemia: Microbiology, clinical features, and significance. *Rev. Infect. Dis.* **1989**, *11*, 208–212. [CrossRef]
44. Harvey, R.L.; Myers, J.P. Nosocomial fungemia in a large community teaching hospital. *Arch. Intern. Med.* **1987**, *147*, 2117–2120. [CrossRef] [PubMed]
45. Verghese, A.; Prabhu, K.; Diamond, R.D.; Sugar, A. Synchronous bacterial and fungal septicemia. A marker for the critically ill surgical patient. *Am. Surg.* **1988**, *54*, 276–283.
46. Tsalidou, M.; Stergiopoulou, T.; Bostanitis, I.; Nikaki, C.; Skoumpa, K.; Koutsoukou, T.; Papaioannidou, P. Surveillance of Antimicrobial Resistance and Multidrug Resistance Prevalence of Clinical Isolates in a Regional Hospital in Northern Greece. *Antibiotics* **2023**, *12*, 1595. [CrossRef] [PubMed]
47. European Centre for Disease Prevention and Control. *Antimicrobial Resistance Surveillance in Europe 2023–2021 Data*; WHO: Geneva, Switzerland, 2023.

48. Yoshida, T.; Tsushima, K.; Tsuchiya, A.; Nishikawa, N.; Shirahata, K.; Kaneko, K.; Ito, K.; Kawakami, H.; Nakagawa, S.; Suzuki, T.; et al. Risk factors for hospital-acquired bacteremia. *Intern. Med.* **2005**, *44*, 1157–1162. [CrossRef]
49. Sun, R.X.; Song, P.; Walline, J.; Wang, H.; Xu, Y.C.; Zhu, H.D.; Yu, X.Z.; Xu, J. Morbidity and mortality risk factors in emergency department patients with *Acinetobacter baumannii* bacteremia. *World J. Emerg. Med.* **2020**, *11*, 164–168. [CrossRef]
50. Crews-Stowe, C.; Sklar, E. The Next Target for Readmission Reporting? Exploring Readmission Rates of Patients with CLABSI. *Antimicrob. Steward. Healthc. Epidemiol.* **2024**, *16*, S71. [CrossRef]
51. Kuo, S.H.; Lin, W.R.; Lin, J.Y.; Huang, C.H.; Jao, Y.T.; Yang, P.W.; Tsai, J.R.; Wang, W.H.; Chen, Y.H.; Hung, C.T.; et al. The epidemiology, antibiograms and predictors of mortality among critically-ill patients with central line-associated bloodstream infections. *J. Microbiol. Immunol. Infect.* **2018**, *51*, 401–410. [CrossRef]
52. Zhou, Q.; Lee, S.K.; Hu, X.J.; Jiang, S.Y.; Chen, C.; Wang, C.Q.; Cao, Y. Successful reduction in central line-associated bloodstream infections in a Chinese neonatal intensive care unit. *Am. J. Infect. Control.* **2015**, *43*, 275–279. [CrossRef]
53. Alwazzeh, M.J.; Alnimr, A.; Al Nassri, S.A.; Alwarthan, S.M.; Alhajri, M.; AlShehail, B.M.; Almubarak, M.; Alghamdi, N.S.; Wali, H.A. Microbiological trends and mortality risk factors of central line-associated bloodstream infections in an academic medical center 2015–2020. *Antimicrob. Resist. Infect. Control* **2023**, *12*, 128. [CrossRef]
54. Rajandra, A.; Yunos, N.M.; Teo, C.H.; Kukreja, A.; Suhaimi, N.A.; Mohd Razali, S.Z.; Basri, S.; Teh, C.S.J.; Leong, C.L.; Ismail, I.; et al. Incidence, Compliance, and Risk Factor Associated with Central Line-Associated Bloodstream Infection (CLABSI) in Intensive Care Unit (ICU) Patients: A Multicenter Study in an Upper Middle-Income Country. *Antibiotics* **2025**, *14*, 271. [CrossRef]
55. Pitiriga, V.; Bakalis, J.; Kampos, E.; Kanellopoulos, P.; Saroglou, G.; Tsakris, A. Duration of central venous catheter placement and central line-associated bloodstream infections after the adoption of prevention bundles: A two-year retrospective study. *Antimicrob. Resist. Infect. Control.* **2022**, *11*, 96. [CrossRef] [PubMed]
56. Burnham, J.P.; Rojek, R.P.; Kollef, M.H. Catheter removal and outcomes of multidrug-resistant central-line-associated bloodstream infection. *Medicine* **2018**, *97*, e12782. [CrossRef] [PubMed]
57. Piatti, G.; Sartini, M.; Cusato, C.; Schito, A.M. Colonization by *Candida auris* in critically ill patients: Role of cutaneous and rectal localization during an outbreak. *J. Hosp. Infect.* **2022**, *120*, 85–89. [CrossRef] [PubMed]
58. Rozwadowski, F.; McAteer, J.; Chow, N.; Skrobarcek, K.; Forsberg, K.; Barrett, P.; Greeley, R.; Fulton, T.; Wells, J.; Welsh, R.; et al. 161. Prevalence and Risk Factors for *Candida auris* Colonization Among Patients in a Long-term Acute Care Hospital—New Jersey, 2017. *Open Forum Infect. Dis.* **2018**, *5*, S14. [CrossRef]
59. Eix, E.F.; Nett, J.E. Modeling *Candida auris* skin colonization: Mice, swine, and humans. *PLoS Pathog.* **2022**, *18*, e1010730. [CrossRef]
60. Eyre, D.W.; Sheppard, A.E.; Madder, H.; Moir, I.; Moroney, R.; Quan, T.P.; Griffiths, D.; George, S.; Butcher, L.; Morgan, M.; et al. A *Candida auris* Outbreak and Its Control in an Intensive Care Setting. *N. Engl. J. Med.* **2018**, *379*, 1322–1331. [CrossRef]
61. Biswal, M.; Rudramurthy, S.M.; Jain, N.; Shamanth, A.S.; Sharma, D.; Jain, K.; Yaddanapudi, L.N.; Chakrabarti, A. Controlling a possible outbreak of *Candida auris* infection: Lessons learnt from multiple interventions. *J. Hosp. Infect.* **2017**, *97*, 363–370. [CrossRef]
62. Rosa, R.; Abbo, L.M.; Jimenez, A.; Carter, C.; Ruiz, M.; Gerald, W.; Jimenez Hamann, M. Effectiveness of a sodium hypochlorite isotonic solution in decolonization of patients with *Candida auris*: Learnings from a county health care system. *Am. J. Infect. Control.* **2024**, *52*, 595–598. [CrossRef]
63. Kilic, A.S.M.; Shi, J.; Green, D.; Wu, F. Molecular and Microbiological Characterization of *Candida auris* Strains Isolated from Colonized and Infected Patients in a New York City Tertiary Care Center. *Am. J. Clin. Pathol.* **2024**, *162*, S177–S178. [CrossRef]
64. Cristina, M.L.; Spagnolo, A.M.; Sartini, M.; Carbone, A.; Oliva, M.; Schinca, E.; Boni, S.; Pontali, E. An Overview on Candida auris in Healthcare Settings. *J. Fungi* **2023**, *9*, 913. [CrossRef] [PubMed]
65. Pandya, N.; Cag, Y.; Pandak, N.; Pekok, A.U.; Poojary, A.; Ayoade, F.; Fasciana, T.; Giammanco, A.; Caskurlu, H.; Rajani, D.P.; et al. International Multicentre Study of *Candida auris* Infections. *J. Fungi* **2021**, *7*, 878. [CrossRef]
66. Frias-De-Leon, M.G.; Hernandez-Castro, R.; Vite-Garin, T.; Arenas, R.; Bonifaz, A.; Castanon-Olivares, L.; Acosta-Altamirano, G.; Martínez-Herrera, E. Antifungal Resistance in *Candida auris*: Molecular Determinants. *Antibiotics* **2020**, *9*, 568. [CrossRef]
67. Rybak, J.M.; Munoz, J.F.; Barker, K.S.; Parker, J.E.; Esquivel, B.D.; Berkow, E.L.; Lockhart, S.R.; Gade, L.; Palmer, G.E.; White, T.C. Mutations in TAC1B: A Novel Genetic Determinant of Clinical Fluconazole Resistance in *Candida auris*. *mBio* **2020**, *11*, e00365-20. [CrossRef] [PubMed]
68. Rybak, J.M.; Barker, K.S.; Munoz, J.F.; Parker, J.E.; Ahmad, S.; Mokaddas, E.; Abdullah, A.; Elhagracy, R.S.; Kelly, S.L.; Cuomo, C.A.; et al. In vivo emergence of high-level resistance during treatment reveals the first identified mechanism of amphotericin B resistance in *Candida auris*. *Clin. Microbiol. Infect.* **2022**, *28*, 838–843. [CrossRef] [PubMed]
69. Lockhart, S.R. *Candida auris* and multidrug resistance: Defining the new normal. *Fungal Genet. Biol.* **2019**, *131*, 103243. [CrossRef]
70. Hager, C.L.; Larkin, E.L.; Long, L.; Zohra Abidi, F.; Shaw, K.J.; Ghannoum, M.A. In Vitro and In Vivo Evaluation of the Antifungal Activity of APX001A/APX001 against *Candida auris*. *Antimicrob. Agents Chemother.* **2018**, *62*, e02319-17. [CrossRef]

71. Ghannoum, M.; Arendrup, M.C.; Chaturvedi, V.P.; Lockhart, S.R.; McCormick, T.S.; Chaturvedi, S.; Berkow, E.L.; Juneja, D.; Tarai, B.; Azie, N.; et al. Ibrexafungerp: A Novel Oral Triterpenoid Antifungal in Development for the Treatment of *Candida auris* Infections. *Antibiotics* **2020**, *9*, 539. [CrossRef]
72. Trevino-Rangel, R.J.; Gonzalez, G.M.; Montoya, A.M.; Rojas, O.C.; Elizondo-Zertuche, M.; Alvarez-Villalobos, N.A. Recent Antifungal Pipeline Developments against *Candida auris*: A Systematic Review. *J. Fungi* **2022**, *8*, 1144. [CrossRef]
73. Ortiz-Roa, C.; Valderrama-Rios, M.C.; Sierra-Umana, S.F.; Rodriguez, J.Y.; Muneton-Lopez, G.A.; Solorzano-Ramos, C.A.; Escandón, P.; Alvarez-Moreno, C.A.; Cortés, J.A. Mortality Caused by *Candida auris* Bloodstream Infections in Comparison with Other Candida Species, a Multicentre Retrospective Cohort. *J. Fungi* **2023**, *9*, 715. [CrossRef] [PubMed]
74. Tabah, A.; Koulenti, D.; Laupland, K.; Misset, B.; Valles, J.; Bruzzi de Carvalho, F.; Paiva, J.A.; Cakar, N.; Ma, X.; Eggimann, P.; et al. Characteristics and determinants of outcome of hospital-acquired bloodstream infections in intensive care units: The EUROBACT International Cohort Study. *Intensive Care Med.* **2012**, *38*, 1930–1945. [CrossRef] [PubMed]
75. Gandra, S.; Tseng, K.K.; Arora, A.; Bhowmik, B.; Robinson, M.L.; Panigrahi, B.; Laxminarayan, R.; Klein, E.Y. The Mortality Burden of Multidrug-resistant Pathogens in India: A Retrospective, Observational Study. *Clin. Infect. Dis.* **2019**, *69*, 563–570. [CrossRef] [PubMed]
76. Schinas, G.; Skintzi, K.; De Lastic, A.-L.; Rodi, M.; Gogos, C.; Mouzaki, A.; Akinosoglou, K. Patterns, Cost, and Immunological Response of MDR vs. Non MDR-Bacteremia: A Prospective Cohort Study. *Pathogens* **2023**, *12*, 1044. [CrossRef]
77. Zasowski, E.J.; Bassetti, M.; Blasi, F.; Goossens, H.; Rello, J.; Sotgiu, G.; Tavoschi, L.; Arber, M.R.; McCool, R.; Patterson, J.V.; et al. A Systematic Review of the Effect of Delayed Appropriate Antibiotic Treatment on the Outcomes of Patients with Severe Bacterial Infections. *Chest* **2020**, *158*, 929–938. [CrossRef]
78. Niederman, M.S. Clinical Impact of Antimicrobial Resistance: Using New Tools to Answer Old Questions. *Chest* **2019**, *155*, 1088–1089. [CrossRef]
79. Bonine, N.G.; Berger, A.; Altincatal, A.; Wang, R.; Bhagnani, T.; Gillard, P.; Lodise, T. Impact of Delayed Appropriate Antibiotic Therapy on Patient Outcomes by Antibiotic Resistance Status from Serious Gram-negative Bacterial Infections. *Am. J. Med. Sci.* **2019**, *357*, 103–110. [CrossRef]
80. Rowan, C.M.; Gertz, S.J.; McArthur, J.; Fitzgerald, J.C.; Nitu, M.E.; Loomis, A.; Hsing, D.D.; Duncan, C.N.; Mahadeo, K.M.; Smith, L.S.; et al. Invasive Mechanical Ventilation and Mortality in Pediatric Hematopoietic Stem Cell Transplantation: A Multicenter Study. *Pediatr. Crit. Care Med.* **2016**, *17*, 294–302. [CrossRef]
81. Gutierrez, L.; Araya, K.; Becerra, M.; Perez, C.; Valenzuela, J.; Lera, L.; Lizana, P.A.; Del Sol, M.; Muñoz-Cofré, R. Predictive value of invasive mechanical ventilation parameters for mortality in COVID-19 related ARDS: A retrospective cohort study. *Sci. Rep.* **2024**, *14*, 13725. [CrossRef]
82. Simonis, F.A.M.; Pelosi, P.; Schultz, M.; Neto, A. Which respiratory and non–respiratory factors contribute to outcome in critically ill patients with invasive ventilation? *Eur. Respir. J.* **2018**, *52*, PA324.
83. Yeung, J.; Couper, K.; Ryan, E.G.; Gates, S.; Hart, N.; Perkins, G.D. Non-invasive ventilation as a strategy for weaning from invasive mechanical ventilation: A systematic review and Bayesian meta-analysis. *Intensive Care Med.* **2018**, *44*, 2192–2204. [CrossRef]
84. Kordalewska, M.; Perlin, D.S. Identification of Drug Resistant *Candida auris*. *Front. Microbiol.* **2019**, *10*, 1918. [CrossRef]

Journal of *Fungi*

MDPI

Review

Pulmonary Aspergillosis in Immunocompromised Critically Ill Patients: Prevalence, Risk Factors, Clinical Features and Diagnosis—A Narrative Review

Maria Grazia Bocci [1,†], Laura Cascarano [2,†], Giulia Capecchi [1,*], Antonio Lesci [3,4], Valerio Sabatini [1], Dorotea Rubino [1], Giulia Valeria Stazi [1], Gabriele Garotto [1], Stefania Carrara [5], Antonella Vulcano [5], Chiara Gori [2], Franca Del Nonno [6], Daniele Colombo [6], Laura Falasca [7], Emanuela Caraffa [8], Stefania Cicalini [8,‡] and Carla Fontana [5,‡]

1 Clinical and Research Department, National Institute for Infectious Diseases "Lazzaro Spallanzani"—IRCCS, 00149 Rome, Italy; mariagrazia.bocci@inmi.it (M.G.B.); valerio.sabatini@inmi.it (V.S.); dorotea.rubino@inmi.it (D.R.); giuliavaleria.stazi@inmi.it (G.V.S.); gabriele.garotto@inmi.it (G.G.)
2 Anesthesia and Intensive Care, San Camillo-Forlanini Hospital, 00152 Rome, Italy; lcascarano@scamilloforlanini.rm.it (L.C.); cgori@scamilloforlanini.rm.it (C.G.)
3 Department of Emergency, Intensive Care Medicine and Anesthesia, Fondazione Policlinico Universitario A. Gemelli IRCCS, 00168 Rome, Italy; antonio.lesci01@icatt.it
4 Istituto di Anestesiologia e Rianimazione, Università Cattolica del Sacro Cuore, 00168 Rome, Italy
5 Laboratory of Microbiology and Biological Bank Unit, National Institute for Infectious Diseases "Lazzaro Spallanzani"—IRCCS, 00149 Rome, Italy; stefania.carrara@inmi.it (S.C.); antonella.vulcano@inmi.it (A.V.); carla.fontana@inmi.it (C.F.)
6 Pathology Unit, National Institute for Infectious Diseases "Lazzaro Spallanzani"—IRCCS, 00149 Roma, Italy; franca.delnonno@inmi.it (F.D.N.); daniele.colombo@inmi.it (D.C.)
7 Laboratory of Electron Microscopy, National Institute for Infectious Diseases "Lazzaro Spallanzani"—IRCCS, 00149 Rome, Italy; laura.falasca@inmi.it
8 Systemic and Immune Depression-Associated Infections Unit, Clinical Department, National Institute for Infectious Diseases "Lazzaro Spallanzani"—IRCCS, 00149 Roma, Italy; emanuela.caraffa@inmi.it (E.C.); stefania.cicalini@inmi.it (S.C.)
* Correspondence: giulia.capecchi@inmi.it
† These authors contributed equally to this work.
‡ These authors contributed equally to this work.

Abstract: Aspergillosis in immunocompromised individuals is a serious and potentially life-threatening infection, as the weakened immune system cannot effectively fight the *Aspergillus* fungus. This review provides an in-depth examination of aspergillosis in patients with various conditions that compromise immunity, including hematological disorders, HIV, SARS-CoV-2 pneumonia, influenza, and those who have undergone solid organ transplants. The clinical manifestations of aspergillosis are influenced by factors such as the host's underlying comorbidities, immune response, and immune suppression due to medications or treatments. The review delves into the epidemiology of aspergillosis, exploring various risk factors that predispose individuals to infection. It also discusses the wide range of clinical symptoms, highlighting the challenges in diagnosis and the importance of early detection. The review contrasts traditional diagnostic approaches with emerging molecular techniques, emphasizing the role of advanced diagnostics in improving outcomes. A proposed clinical decision-making flowchart is provided to assist healthcare professionals in managing suspected cases of aspergillosis. In addition to diagnostic challenges, the review addresses antifungal prophylaxis, pre-emptive therapy, and the growing concern of pharmacological resistance to antifungal agents. It concludes with a discussion of future research directions, underscoring the need for improved therapeutic strategies and preventative measures in immunocompromised patients to reduce the burden of this severe fungal infection.

Keywords: *Aspergillus*; pulmonary aspergillosis; invasive aspergillosis; immunocompromised patients; antifungal treatment

1. Introduction

Aspergillosis is an opportunistic fungal infection caused by molds of the genus *Aspergillus*, characterized by a wide spectrum of clinical manifestations. These range from allergic forms, such as allergic bronchopulmonary aspergillosis (ABPA) and severe asthma with fungal sensitization (SAFS), to progressive chronic forms like chronic pulmonary aspergillosis (CPA), and finally to acute, potentially lethal forms of invasive aspergillosis (IA) [1] (Figure 1). In recent years, novel clinical entities have emerged in association with severe viral infections, including influenza-associated pulmonary aspergillosis (IAPA) and COVID-19—associated pulmonary aspergillosis (CAPA), which have been described in critically ill patients admitted to intensive care units, even in the absence of classical immunosuppressive risk factors [2,3]. From an etiological standpoint, *Aspergillus fumigatus* remains the principal pathogen responsible for aspergillosis [1,4]. Other relevant species include *Aspergillus flavus*, *Aspergillus niger*, *Aspergillus terreus*, *Aspergillus nidulans*, and *Aspergillus calidoustus*, which are among the most frequently involved in human disease [1]. These names actually refer to complexes of cryptic species that are morphologically similar [1]. Although *A. fumigatus* sensu stricto is by far the most prevalent, cryptic related species (e.g., *Aspergillus lentulus*, *Aspergillus udagawae*, *Aspergillus viridinutans*, and *Aspergillus calidoustus*) are increasingly recognized and may account for up to 10–30% of IA cases [1,5]. These emerging species are clinically relevant due to intrinsic antifungal resistance and diagnostic challenges, often associated with refractory or relapsing IA [1]. From both epidemiological and clinical perspectives, aspergillosis plays a major role in immunocompromised populations. Patients with hematologic malignancies (especially acute leukemias) undergoing intensive chemotherapy or hematopoietic stem cell transplantation (HSCT), as well as solid organ transplant recipients, individuals with advanced HIV infection, or those receiving high-dose corticosteroids or novel immunosuppressive agents (e.g., targeted biologics), are at particularly high risk for IA [6]. In these groups, *Aspergillus* remains one of the leading causes of infectious mortality [1].

Importantly, severe viral respiratory infections such as influenza and COVID-19 can predispose even non-classically immunocompromised individuals to IA, particularly those with acute respiratory distress syndrome (ARDS). In such cases, IAPA and CAPA have been linked to poor outcomes, with mortality rates often exceeding 50% [2,3]. This evolving risk landscape has prompted updates in diagnostic definitions and prevention strategies. In 2020, the European Organization for Research and Treatment of Cancer and the Mycoses Study Group (EORTC/MSG) consortium revised the definitions of invasive fungal disease (IFD), expanding the list of host factors (e.g., solid organ transplantation, certain inborn errors of immunity such as STAT3 deficiency) and introducing new microbiological criteria (such as PCR testing on clinical specimens) to better identify probable aspergillosis cases [6,7]. These updated definitions [7] offer a broader framework to recognize IA beyond the classical population of hematologic patients, acknowledging that IA is now increasingly reported in critically ill patients without prolonged neutropenia (e.g., ICU patients with COVID-19) [8]. In parallel, guidelines have been developed for newly identified at-risk populations. For instance, specific diagnostic criteria for CAPA have been proposed during the pandemic [9], and antifungal prophylaxis recommendations have been formulated for patients with acute myeloid leukemia (AML) receiving novel targeted therapies, given the persistent high risk of IFD in this context [10].

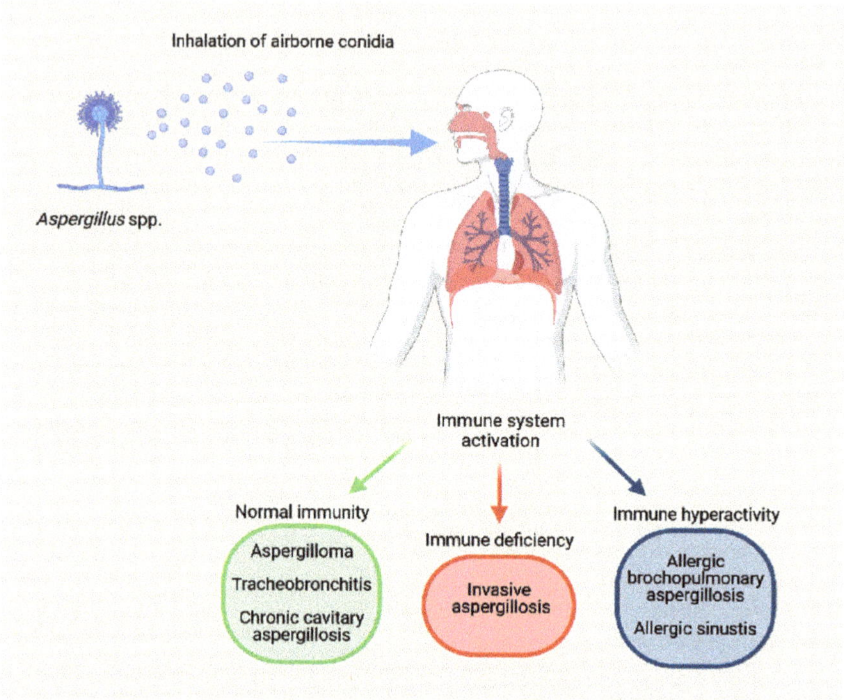

Figure 1. Spectrum of disease due to interaction between *Aspergillus* and the human host. The inhalation of *Aspergillus* spp. airborne conidia can cause different categories of human disease depending on the immunological status of the infected host. Among immunocompetent individuals, diseases such as aspergilloma, tracheobronchitis, or chronic cavitary pulmonary aspergillosis (CCPA) are typically caused by *A. fumigatus*, but other species such as *A. niger*, *A. flavus*, and *A. terreus* may also be involved. In immunocompromised patients, IA is predominantly associated with *A. fumigatus*, but emerging pathogens like *A. terreus*, *A. nidulans*, *A. flavus*, and cryptic species in the *A. fumigatus* complex (e.g., *A. lentulus*, *A. udagawae*) have increasingly been reported, often associated with intrinsic antifungal resistance. Conversely, patients with immune system hyperactivation may develop allergic manifestations such as severe asthma with fungal sensitization (SAFS), allergic bronchopulmonary aspergillosis (ABPA), and allergic fungal rhinosinusitis (AFRS), mostly related to *A. fumigatus* and occasionally *A. niger*.

Here, we will specifically address three major categories of immunocompromised hosts: people living with HIV, solid organ transplant recipients, and patients with hematologic malignancies or post-HSCT. In addition, we will examine viral-associated pulmonary aspergillosis, with a focus on influenza-associated pulmonary aspergillosis (IAPA) and COVID-19-associated pulmonary aspergillosis (CAPA), which have emerged as distinct clinical entities. Although these forms can also occur in patients with classical immunosuppression, they have been increasingly reported in critically ill patients without traditional risk factors, suggesting a broader at-risk population [11].

2. Microbiology, Transmission and Pathogenesis

Species in the filamentous fungal genus *Aspergillus* display a wide diversity of lifestyles and are of great importance to humans [12]. On the other hand, several *Aspergillus* species are known to cause human disease, including *A. fumigatus*, *A. flavus*, *A. niger*, *A. terreus*, *A. versicolor*, *A. ustus*, *A. lentulus*, and *A. nidulans* (also known as *Emericella nidulans*). These molds are ubiquitous in the environment and produce airborne conidia [13]. *Aspergillus* species differ not only in pathogenicity but also in morphological characteristics, including the size of their conidia, which may influence airborne dispersal and deposition in the respiratory tract.

The conidial size of *A. fumigatus*, typically ranges from 2.0 to 3.5 μm, favoring deep alveolar penetration [14,15]. In contrast, *A. flavus* conidia are generally 3.0–6.0 μm in diameter, slightly larger and more irregular in shape [16]; *A. niger* and *A. terreus* produce conidia of approximately 3.5–5.0 μm, which may deposit higher in the airways and cryptic species such as *A. lentulus* and *A. calidoustus* often resemble *A. fumigatus* macroscopically, but have slightly larger or more variable conidia (2.5–4.5 μm), requiring molecular identification for accurate differentiation [17]. These variations in size can affect the aerobiology and infectious potential of each species, with smaller conidia better suited to reach the terminal alveoli—a key factor in the pathogenesis of invasive pulmonary aspergillosis (IPA) [18]. Humans inhale hundreds to thousands of these infectious propagules on a daily basis. Due to their small size, *Aspergillus* spp. conidia can bypass mucociliary clearance and lodge in the lower respiratory tract. In immunocompetent individuals, the coordinated action of the respiratory epithelium, lung-resident macrophages, as well as recruited neutrophils and monocytes clear conidia efficiently [19]. The removal of fungi from an infected host depends on the rapid migration of a sufficient number of phagocytes to the site of infection and the recognition of fungal PAMPs via PRRs (both soluble and membrane-bound). This leads to the ingestion and, ultimately, the degradation of the ingested fungal cells within the phagocytes [20]. A key aspect of this process is the maturation of conidia, which, upon swelling, lose the thin hydrophobic RodA protein layer, a surface component of *Aspergillus* that masks the immunogenic components of the cell wall, such as β-glucan. This allows for increased exposure of fungal PAMPs, facilitating recognition by PRRs such as Dectin-1 and Dectin-2. Dectin-1, expressed on macrophages, neutrophils, and dendritic cells, recognizes β-glucan fragments on swollen conidia, mediating a prolonged inflammatory response and the activation of immune cells. Similarly, Dectin-2, expressed on macrophages and dendritic cells, binds fungal mannans, facilitating early recognition of conidia and the subsequent recruitment of phagocytes [21].

CR3 (CD11b/CD18), a pattern recognition receptor expressed on phagocytes, binds fungal chitosan and plays a critical role in orchestrating both phagocytosis and cytokine release in response to *Aspergillus* [22]. A key molecule in this early recognition phase is pentraxin 3 (PTX3), which enhances phagocytosis by binding to fungal surfaces [23]. Activated immune cells then release chemokines such as CXCL1 and CXCL5, which recruit neutrophils to the site of infection. Other inflammatory mediators—such as leukotriene B4 (LTB4), complement component C5a, and galectin-3—further amplify neutrophil recruitment and promote swarming behavior [24]. In addition to degranulation, phagocytosis, ROS generation, and cytokine production, pathogen-induced activation of neutrophils also initiates cellular processes to expel chromatin to the exterior for neutrophil extracellular trap (NET) release. NETs also appear to modulate host immunity to *A. fumigatus* through release of long pentraxin (PTX) 3 facilitating pathogen recognition [25,26].

Monocytes migrate into the alveolar space, where they differentiate into monocyte-derived dendritic cells and collaborate with neutrophils to engulf *Aspergillus* conidia.

Within the phagolysosomes, fungal killing is driven by reactive oxygen species (ROS) generated by NADPH oxidase. The subsequent release of interleukins and other pro-inflammatory mediators further amplifies the immune response by recruiting plasmacytoid dendritic cells. These cells enhance the oxidative burst, promoting efficient conidial clearance from the lungs [27,28].

Cell-mediated immunity plays a crucial role in antifungal immunity or allergy-associated tissue damage. Fungal antigens are carried to lymph nodes, where they are processed and presented by antigen-presenting cells, leading to the differentiation of naive $CD4^+$ T cells into distinct T-helper subsets. In the context of invasive aspergillosis, the functional polarization of these T cells is critical in determining the outcome of infection. Th1 cells, characterized by T-bet expression and the production of interferon-gamma, promote macrophage activation [29]. A robust Th1 response is generally associated with protective immunity, and its deficiency correlates with increased susceptibility to IA.

For the Th17 cells, the differentiation is driven by innate cytokines such as IL-1α, IL-1β, IL-6, and IL-23, which activate the lineage-defining transcription factor RORγT (retinoic acid receptor-related orphan receptor gamma T). These Th17 cells secrete IL-17A and IL-22, promoting neutrophil recruitment and strengthening epithelial barrier defenses through the induction of antimicrobial peptides. While Th17 responses contribute to fungal control in the early phases of infection, excessive or prolonged activation may lead to immunopathology and tissue damage. Th2 responses, characterized by IL-4, IL-5, and IL-13 production, are considered detrimental in IA, as they interfere with protective Th1 activity and may impair antifungal effector mechanisms.

Regulatory T cells (Tregs), defined by Foxp3 expression and the secretion of IL-10 and TGF-β, act to restrain excessive immune activation and limit tissue damage. However, in the setting of IA, an overabundance or overactivity of Tregs may suppress effective Th1 and Th17 responses, contributing to fungal persistence. These observations highlight the necessity of a balanced adaptive immune response in IA—one that ensures pathogen clearance while preventing immune-mediated injury [30,31]. Fungal antigens are carried to lymph nodes, where T helper cells differentiate into Tbet+ T helper 1 cells. These cells then travel to the lungs to produce IFN-γ, activating macrophages antifungal functions. Meanwhile, T helper 2 cells, which secrete IL-4, IL-5, and IL-13, are implicated in allergic responses. Additionally, T helper 17 cells, through their secretion of IL-17, contribute to certain antifungal activities [29].

3. Immunocompromised Patients

3.1. Hematologic Patients

The survival rates of patients with hematological malignancies have seen significant improvements due to advancements in medication and supportive therapies, which have also led to a rise in ICU admissions [32].

These patients are more susceptible to developing serious complications, one of which is invasive fungal infections (IFI) [33]. IFIs are common in bone marrow transplantation (BMT), with an incidence ranging from 5% to 20%. Risk factors include the development of graft-versus-host disease, increased human leukocyte antigen mismatch, and co-infection with respiratory viruses or cytomegalovirus (CMV). IA represents the most prevalent IFI among recipients, accounting for 43% of cases, with *A. fumigatus* being the predominant species [34].

The diagnosis of IFI in hematology patients is associated with the highest mortality [35].

In Europe, 3.7% of ICU patients had proven or probable IA with an ICU mortality of 87–97% in immunocompromised patients [36,37] while the incidence among patients with leukemia is almost 11.7%, but the use of antifungal prophylaxis in high-risk patients has modified this data [38,39].

All blood malignancies involve the immune system, which promotes the IFI development for T cell dysfunction, hypogammaglobinemia, neutropenia due to bone marrow involvement, and leukopenia due to chemotherapy or BMT.

As seen before, neutrophils are the first defense line against fungi: an absolute neutrophil count < 500/mm^3 predisposes to infections as much as an average count but immature neutrophils (acute myeloid leukemia) [40,41]. The most critical risk factor for IA is prolonged and profound neutropenia. Neutropenia is frequent in patients with acute leukemia or myelodysplastic syndrome (MDS) during remission induction chemotherapy and in patients undergoing allogeneic HSCT. In the former, the period more susceptible to IFI is the time until the first remission. In contrast, the next high-risk period in the latter group follows the conditioning chemotherapy regimen.

A deficiency in lymphocyte production and trafficking and alterations in lymph organ function, second line of defense against fungi, promote IFI in lymphocytic leukemia involving innate and adaptive immunity [42,43].

In HSCT patients also, the type of immunosuppression can promote an impairment of the host's defense, and after neutropenia resolution, the development of a graft-versus-host disease (GvHD) may require escalation of immunosuppression therapy, increasing the risk of IFI. Other significant risk factors include renal or liver dysfunction as well as age over 50 years, and they represent the strongest mortality predictors among patients with acute myelogenous leukemia (AML) post induction chemotherapy [44].

Signs or symptoms are not so specific in an early phase. The challenge in the management of IA in hematologic patients is to early identify high-risk patients through host risk factors, clinical criteria and microbiological criteria to make a presumed diagnosis to start therapy and reduce mortality. A defined diagnosis can be made through autopsy or with a positive culture of *Aspergillus* in a specimen obtained by biopsy or needle aspiration from a sterile site that is clinically or radiologically abnormal in which hyphae or melanized yeast-like forms are present.

The gold standard for diagnosing IA remains culture, as it allows precise identification of the *Aspergillus* species, thanks to advancements in identification techniques, such as the introduction of MALDI-TOF assay, as well as antimicrobial susceptibility testing of the isolate. However, the diagnosis of IA is still challenging for several reasons: tissue sampling may be difficult or contraindicated in patients with hemodynamic instability, thrombocytopenia, or coagulation disorders; the sensitivity of cultures is often low and may vary depending on the type and quality of the specimen, the timing of collection, and the prior use of antifungal agents [45,46].

For this reason, indirect methods must be introduced, such as detecting galactomannan (GM) or β-D-glucan (BDG) in serum, bronchoalveolar lavage (BAL), or polymerase chain reaction (PCR) assay [7].

A recent systematic review by Bukkems et al. [47] describes GM antigen assay role in invasive pulmonary aspergillosis (IPA) diagnosis in hematological patients, including all IPA categories. Considering the type of patients, the type of sample analyzed, and the IPA categories is essential to establish the sensitivity and specificity of GM assays. The results show that for serum and BAL, an optical density index (ODI) of 0.5 and 1.0, respectively, provides the best diagnostic accuracy with a sensitivity and specificity different for each IPA category (Table 1).

Table 1. Reliability of GM test with ODI 0.5 for serum/plasma and 1.0 for BAL.

Sample/Test	Sensitivity	Specificity
Serum/plasma (proven/probable vs. no IPA)	76%	92%
Serum/plasma (proven/probable/possible vs. no IPA)	45%	91%
BAL (proven/probable vs. no IPA)	80%	95%
BAL (proven/probable/possible vs. no IPA)	49%	95%

Without treatment, mortality for IA is 100% in these patients, while with antifungal treatment the overall mortality is between 30% and 40% for AML and 60% for HSCT patients [48].

3.2. HIV Patients

In the absence of treatment, HIV infection can cause marked deterioration of immune function. The degree of immune suppression is directly linked to the emergence of opportunistic infections, particularly when circulating CD4+ lymphocytes are below 200 cells/mm^3. Among these infections, fungal diseases such as pneumocystosis, cryptococcosis, histoplasmosis, and aspergillosis contribute significantly to morbidity and mortality in individuals with AIDS [49].

Fungi belonging to the *Aspergillus* genus are responsible for a range of clinical syndromes with varying degrees of severity in humans [50]. Individuals with advanced stages of HIV are vulnerable to different manifestations of aspergillosis, including invasive forms, atypical bronchial obstructions, and chronic pulmonary involvement. Although the lungs represent the primary site of infection, dissemination to other organs such as the heart, paranasal sinuses, kidneys, and central nervous system can also occur [51]. Superficial and allergic forms, such as nail infections, keratitis, and outer ear involvement, are relatively uncommon in this population, although some studies suggest keratitis may be slightly more frequent in HIV-infected individuals [52].

In contrast to patients with neutropenia or those undergoing transplantation, in whom the disease typically presents acutely, people with HIV (PWH) often experience more protracted, subacute forms of infection. Nevertheless, both the acute and chronic presentations may carry a significant risk of death [53]. The frequency of aspergillosis in PWH has declined since the introduction of effective antiretroviral therapy, but the infection still occurs [54,55]. An analysis of over 35,000 patients from a U.S. national database reported an incidence of 3.5 cases per 1000 person-years [56], consistent with other large-scale studies reporting incidence rates around 0.43% [54,55]. Neutropenia and corticosteroid therapy were present in nearly half of all reported cases [53,55], and other contributing factors include concurrent *Pneumocystis jirovecii* infection and CD4+ counts below 50–100 cells/mm^3 [57–63].

In their review, Doumbo et al. [64] described the epidemiological context in Mali, where health systems are under strain due to high tuberculosis prevalence and a moderate HIV burden. The estimated population of Mali was 21,251,000 in 2020 (UN), of which 45% were children < 14 years of age. Among HIV-positive individuals, the less frequently reported invasive fungal infections included IA (n = 1230), keratitis (n = 2820), bloodstream *Candida* infections (>1060), and mucormycosis (n = 43). These numbers are thought to underestimate the true burden of fungal infections due to insufficient access to diagnostic resources.

A study by Iruda et al. [65], involving 25 HIV-positive patients (aged 23–58 years), identified 11 cases of invasive and 14 cases of chronic pulmonary aspergillosis. The overall prevalence in their cohort of nearly 20,000 individuals was 0.1%. Most patients with the invasive form had CD4+ counts below 100 cells/mm^3, whereas over 40% of those with chronic disease had counts above 200 cells/mm^3. A history of tuberculosis was common, particularly among patients with chronic aspergillosis (85.7%). The diagnosis was generally supported by radiological findings and culture positivity (71.4%). Crude mortality was high: 72.7% in the invasive group and 42.8% in the chronic group. Despite its rarity (0.1% in PWH), IA should be suspected in immunocompromised patients with unexplained pneumonia refractory to conventional therapy, and chronic pulmonary disease should be considered in those with respiratory decline and a previous history of tuberculosis.

In a systematic review, Yerbanga IW et al. [66] confirmed the presence of IA in Africa, highlighting a substantial number of undetected cases. The review emphasized key differences in epidemiology between African countries and high-income settings and called for more targeted surveillance in regions with growing numbers of at-risk individuals.

Denning et al. [67] analyzed 54 studies and identified 859 HIV/AIDS patients with aspergillosis. Among the 853 with available outcome data, 707 had died, corresponding to a mortality rate of 83%. The majority of reported cases over the past 15 years occurred in untreated individuals, those with poor adherence to antiretroviral therapy, or in advanced stages of immunosuppression, particularly when CD4+ counts were under 50 cells/mm^3. Although aspergillosis is no longer listed among AIDS-defining conditions, its high fatality rate in this population underscores the importance of considering it in patients presenting with fever of unknown origin and/or lung infiltrates, particularly when cavitary lesions are present.

3.3. Organ Recipients

Solid organ transplantation (SOT) has significantly improved the lives of individuals with end-stage organ failure. Despite the success, SOT recipients are at a higher risk of opportunistic infections due to long-term immunosuppression, with IA being a notable cause of morbidity and mortality. The management of IA in SOT recipients remains challenging despite advancements in transplant medicine and antifungal therapies.

A study conducted on a cohort of 960 cases with probable or proven IA, as reported in the Prospective Antifungal Therapy Alliance (PATH Alliance) registry, revealed that 48.3% of the individuals had hematologic malignancies, 29.2% were recipients of SOT, 27.9% had undergone HSCT, and 33.8% were neutropenic. Among these patients, the prevalent clinical manifestations of IA included IPA and rhinocerebral aspergillosis [68].

Despite a decline in mortality rates among transplant recipients in recent years from 65–92% to 22%, it is estimated that approximately 9.3–16.9% of all deaths occurring within the first-year post-transplantation can still be attributed to IA [69].

3.4. Liver Transplant Recipients

Liver transplant (LT) recipients are at risk for IA, with a prevalence of 1.8%, according to recent studies [68,70]. This population's mortality rate associated with IA ranges from 50% to 100%. However, there has been a significant decrease in both the incidence and mortality rates over the past decades, with the incidence dropping from 10% to 1.8% and mortality declining from 100% to approximately 50% [68,70,71].

Despite these improvements, the mortality rates among LT recipients remain higher compared to other transplant procedures [72]. IA manifests earlier in LT recipients than in other SOT recipients, excluding the heart. The diagnosis is typically made within 17 to

128 days post-transplantation, whereas for all SOT recipients the median time to diagnosis is around 100 days [69,73–75].

Risk factors for IA in LT recipients include immune system dysregulation induced by bacterial translocation, complement system insufficiency, monocyte suppression, and impaired neutrophil phagocytosis. Management of LT recipients involves reduced levels of immunosuppression, with some individuals potentially achieving immunological tolerance. Corticosteroids have been identified as a risk factor for IA, while calcineurin inhibitors exhibit activities against *Aspergillus* species [76–78].

Other risk factors for IA in LT recipients include pre-transplant hepatic failure, primary allograft failure or dysfunction, re-transplantation, post-transplant dialysis, and elevated transfusion needs. CMV infection has been associated with a higher likelihood of IA occurring beyond 100 days post-transplantation, and a MELD score exceeding 30 is linked to a heightened risk of developing IFIs [73,79–81]. Steroid boluses administered before surgery were linked to an increased occurrence of IPA after transplant operations, with a rate of 16% versus a mere 0.4% [70]. Additionally, a higher incidence of IFIs was observed in patients who received corticosteroids or antibiotics within three months before surgery, showing a significant difference (64% vs. 34% for corticosteroids and 61% vs. 31% for antibiotics) [82]. Factors such as surgical durations longer than 11 h, extensive red blood cell transfusions during surgery (with an odds ratio of 23.1), and plasma and platelet transfusions were strongly associated with an increased risk of IFIs. Surgical procedures such as [83–85] choledochojejunostomy and Roux-en-Y anastomosis were also associated with a significantly higher risk of IFIs, two-fold and 16-fold, respectively [69,86]. In a comprehensive study of 1730 patients undergoing LT, with subsequent development of CNS lesions, 11 were found to have lesions due to infections. *A. fumigatus* was the most common agent, typically appearing about 21 days after transplantation, with cases ranging from 10 to 27 days post-operation [87].

3.5. Renal Transplant Recipients

The occurrence of IA in individuals who have received a kidney transplant (KT) is reported to be between 0.7% and 4% [88,89]. Factors that elevate the likelihood of IA within this group encompass the use of high doses of methylprednisolone, prolonged therapy with corticosteroids, and the necessity for hemodialysis due to transplant failure [90].

Following SOT, particularly in the initial six months, there is a heightened use of drugs that suppress the immune system. This period is notably vulnerable to opportunistic infections, including IPA [91]. Despite this, recipients of KTs typically undergo a continuous treatment regimen that includes corticosteroids, calcineurin inhibitors, and anti-proliferative agents [92]. Although KT recipients have a comparatively lower incidence of post-transplant IPA than recipients of other transplants, the extension of KT survival has increased in older patients undergoing prolonged intensive immunosuppressive treatment [93].

Although the risk of post-transplant IPA is lower in KT recipients compared to other transplant recipients, advances in the longevity of KT recipients have led to an increasing number of elderly patients receiving prolonged high levels of immunosuppression [94]. It has also been documented that significant risk factors for the development of IA after KT include the receipt of methylprednisolone in doses greater than 3 g, prolonged use of corticosteroids, and the need for hemodialysis following transplant failure [95,96].

3.6. Lung Transplant Recipients

In recipients of lung transplant (LT), IA emerges as the most common fungal infection. Historically, the occurrence of this infection in LT patients was estimated to be between 4%

and 23%. However, recent research indicates a decline in its prevalence [97]. The time from LT to the development of IA has increased, a change attributed to the use of antifungal prophylaxis early after transplantation. This has led to a shift in the median time of onset of the infection from 120 days to 483 and 508 days post-transplant, respectively [98].

A. fumigatus is the primary fungal species responsible for IA in these patients. A study tracking 900 adult LT recipients in various international centers for four years found that 79 individuals had 115 cases of infection. Another study of 251 patients showed that *Aspergillus* was present in 86 (33%) cases, ranging from colonization to tracheobronchitis and IA [98].

The significance of *Aspergillus* spp. colonization before LT remains controversial. Some researchers emphasize its importance, particularly during the first-year post-transplant and when colonization occurs within six months prior to surgery. However, other studies do not support a clear link between pre-transplant colonization and post-transplant IA [99,100].

Other identified risk factors for IA in LT recipients include bronchial anastomotic leaks, surgical site complications, airway narrowing, allograft dysfunction or ischemia, reperfusion injury, CMV infection, bronchiolitis, and the need for increased immunosuppression to prevent graft rejection [100].

Ischemic damage at the bronchial anastomotic site can lead to ulcerative tracheobronchitis, a form of IA, and potentially bronchovascular fistulas, which are associated with serious bleeding risks [69,100]. Mortality rates for LT recipients with IA vary significantly, from 23% to 29% in tracheobronchitis cases to 67–82% in severe pulmonary infections. Nonetheless, some data suggest the mortality rate could be as low as 20%. Patients with central nervous system involvement or disseminated disease generally have a poorer prognosis [87].

3.7. Heart Transplant Recipients

The incidence of IA among heart transplant (HT) recipients is reported to be between 1% and 14%, with the risks for developing this condition differing based on whether it occurs early or later after the transplant. Factors such as undergoing hemodialysis, needing thoracic surgery again, and increased levels of immunosuppressive drugs are linked to a higher likelihood of developing IA among these patients. Not all patients with IA will show signs of neutropenia, making it essential to diagnose the condition through clinical observation and imaging results. The most common presentation of IA in this group is IPA, particularly within the first three months following a HT. On the other hand, cases of IA that arise later tend to show a wider spread of the disease, including to the CNS and other areas outside the lungs [100–102]. Research by Montoya and colleagues revealed that HT patients with either form of aspergillosis did not have neutropenia. This underscores the importance of considering IPA in patients who, despite not having neutropenia, exhibit fever and respiratory symptoms, have positive cultures from respiratory secretions, and show unusual patterns on lung scans, such as nodules, especially within three months after their transplant [103].

A 2006 study highlighted a one-year post-diagnosis mortality rate of 67% among HT patients with IA [104]. However, more recent data suggest a reduction in mortality to 38%, indicating progress in managing this condition in the transplant population [102].

4. Coinfections

4.1. COVID-Associated Pulmonary Aspergillosis (CAPA)

Historically, the development of pulmonary aspergillosis was mainly linked to specific conditions of profound immunosuppression, such as hematologic malignancies, stem cell transplantation, cytotoxic therapies, or prolonged use of corticosteroids and other immunosuppressants. More recently, the spectrum of risk factors has widened to include inherited immunodeficiencies and reduced lymphocyte subsets [58,105].

The COVID-19 pandemic has introduced new clinical scenarios. Infection with SARS-CoV-2 is now recognized to induce complex immune dysregulation, affecting both T helper 1 and 2 responses [106–108]. Although a direct inhibitory effect on antifungal host defense has not been definitively proven [109], there is growing concern that either the infection itself or the immunomodulatory treatments commonly used—such as corticosteroids and biological agents—may facilitate opportunistic fungal infections. This has prompted the identification of a specific form of aspergillosis emerging in COVID-19 patients, termed COVID-19-associated pulmonary aspergillosis (CAPA) [9]. However, evidence on the prevalence and relative contribution of different risk factors in this setting remains limited. Notably, in patients with COVID-19-associated acute respiratory distress syndrome (ARDS) and concomitant severe infections, mortality has been reported to increase by 16–25% when aspergillosis is present [110,111].

Montrucchio et al. highlighted the relevance of prolonged respiratory support, both invasive and non-invasive, which is commonly required in severe COVID-19, as a potential risk factor for CAPA [112]. Additional attention should be given to the possible role of corticosteroids and immunobiological therapies in increasing both incidence and mortality. Furthermore, the endothelial damage caused by SARS-CoV-2 [113] and its effects on the host immune response [114] may contribute to increased susceptibility to fungal infections.

Environmental and logistical challenges in ICUs during the COVID-19 pandemic—such as high patient volume, altered ventilation systems, structural renovations, and extended use of isolation rooms—might also have played a role in promoting fungal colonization and infection.

Chavda et al. reported a high prevalence of opportunistic fungal infections among COVID-19 patients, particularly in the elderly population [115]. Data from the UK indicate that individuals aged between 55 and 81 years represent the highest proportion of coinfected patients [116], while a Spanish study found the median age to be 62 years (range 48–74) [117]. In a retrospective cohort from a tertiary care center in North India, Paul et al. documented that COVID-19 patients with diabetes mellitus were at significantly increased risk of developing angio-invasive fungal infections [118]. A prospective study conducted at the India Institute of Medical Sciences (AIIMS), in New Delhi, on ten patients with subacute IPA, all with diabetes and steroid exposure, described cough—often preceded by hemoptysis—as a typical symptom [119].

The diagnosis of CAPA remains challenging, primarily due to its clinical and radiological resemblance to severe COVID-19 pneumonia. Moreover, conventional diagnostic markers such as serum galactomannan may show reduced sensitivity in non-neutropenic patients, as neutrophils can clear fungal antigens from the bloodstream before detection [120]. This has led to a disease course that, unlike classical angio-invasive aspergillosis in neutropenic patients, is often initially limited to airway invasion for several days before vascular dissemination occurs.

Two years after the onset of the pandemic, numerous epidemiological and observational studies continue to emphasize the clinical impact of SARS-CoV-2 in predisposing patients to secondary fungal infections. The emergence of these coinfections appears to be strongly influenced by the use of mechanical ventilation, intravascular catheters, and immunosuppressive therapies [121].

4.2. Influenza-Associated Pulmonary Aspergillosis (IAPA)

Influenza-associated pulmonary aspergillosis (IAPA) is an emerging complication of influenza infection, often associated with *Aspergillus* tracheobronchitis [122,123]. It has emerged as a significant complication in critically ill patients, particularly those admitted to ICUs with severe influenza, markedly increasing influenza-associated mortality. Traditionally considered a concern limited to severely immunocompromised hosts, recent evidence shows that IAPA can occur in non-neutropenic patients without classical risk factors, expanding the at-risk population [11,124]. Recent cohort studies and meta-analyses estimate the incidence of IAPA in ICU influenza patients to range between 10% and 20%, with mortality rates frequently exceeding 50% [2,11,122,125]. In a large retrospective multicenter cohort by Schauwvlieghe et al., 19% of influenza ICU patients developed IAPA, with a 51% mortality rate compared to 28% in those without aspergillosis [11]. Another recent Swiss study confirmed a prevalence of ~11%, with early onset (within 72 h of ICU admission), highlighting the aggressive and rapidly progressive nature of IAPA [124,125]. Pathophysiologically, influenza virus causes extensive damage to the respiratory epithelium, disrupts mucociliary clearance, and impairs local immune responses, creating a permissive environment for *Aspergillus* invasion. In addition, the viral infection induces dysregulation of innate immunity, particularly of neutrophil and macrophage function, while promoting excessive interferon-γ responses, which paradoxically impair antifungal activity, resulting in a defective Th17-immune response, depletion of macrophages, and impaired killing of *Aspergillus* conidia by macrophages [126]. However, recent experimental evidence suggests that restoration of IL-17 signaling alone is insufficient to ensure adequate fungal clearance during IAPA, indicating the involvement of additional immunological impairments contributing to defective antifungal immunity [126]. Supporting this hypothesis, Garcia et al. demonstrated in a murine model of IAPA that inhaled immunotherapy with pattern recognition receptor (PRR) agonists effectively reactivated innate immune responses, reversed profound immune paralysis, and improved *Aspergillus* clearance and survival [127,128]. A growing body of evidence identifies asthma and inhaled corticosteroid use as independent risk factors for IAPA, even in the absence of systemic immunosuppression [125]. Chronic pulmonary disease (e.g., COPD), male sex, and systemic corticosteroids further increase susceptibility [122,125]. Diagnosis remains challenging, as clinical and radiological findings of IAPA often overlap with those of viral pneumonia. Bronchoscopy with bronchoalveolar lavage (BAL) is recommended when feasible. Galactomannan (GM) detection in BAL fluid shows higher sensitivity than serum GM in IAPA patients [122]. Fungal culture, PCR, and β-D-glucan may provide supportive evidence. Management includes early antifungal therapy, most commonly with voriconazole or isavuconazole. Empirical antifungal treatment is often warranted in high-risk patients with persistent or worsening respiratory failure. However, data on prophylaxis remain inconclusive: the POSA-FLU trial assessing posaconazole prophylaxis in influenza ICU patients showed limited benefit, suggesting the need for better risk stratification [129]. Outcomes remain poor, especially when diagnosis is delayed or antifungal therapy is not promptly initiated. Mortality rates are significantly higher in IAPA patients compared to influenza patients without fungal co-infection, underscoring the need for heightened clinical awareness, early diagnostic workup, and multidisciplinary management.

5. Diagnosis

Because of the paucity of clinical signs and the limited sensitivity and/or specificity of radiology and mycological tests, the diagnosis of IPA is graded according to a scale of probability (possible, probable or proven) of disease [46,130]. The first version of definitions of invasive fungal infections (IFIs) for clinical and epidemiological research published in 2002 by a consensus group of the EORTC/MSG, refer to immunocompromised patients with cancer and HSCT [131]. In 2008, they had an update [132] in which data were insufficient to establish a threshold for *Aspergillus* GM, and the diagnostic role of 1,3-beta-D-glucan (BDG) was uncertain.

It is pertinent to mention that the updated 2019 European Organization for Research and Treatment of Cancer/Mycoses Study Group Education and Research Consortium (EORTC/MSGERC) consensus definitions for invasive fungal disease (IFD) now include: (i) *Aspergillus* PCR and (ii) the molecular detection of fungal (*Aspergillus*) DNA in tissue, as diagnostic criteria for IA [45].

The 2020 Committee's update tried to incorporate all these new findings [7,133,134] (Figure 2).

For IFI, as well as for aspergillosis, the definitions assigned three levels of probability to the diagnosis: for proven infection, it is necessary to confirm the presence of fungi in a sterile material during the microscopic exam with alterations in tissue structure as illustrated in Figure 3.

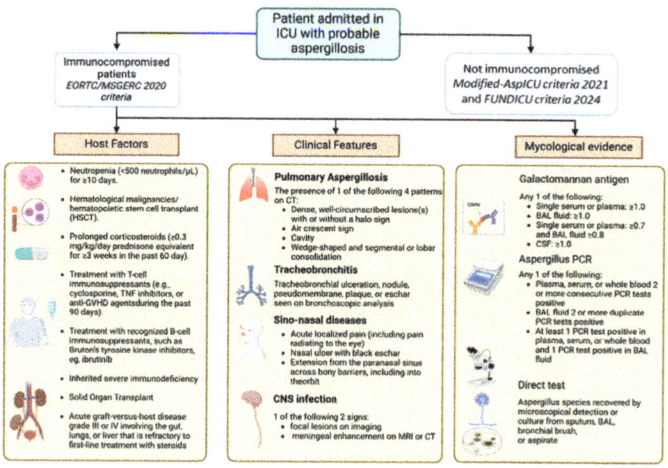

Figure 2. Criteria for probable aspergillosis diagnosis in patients admitted in ICU. Various criteria have been proposed for diagnosing probable aspergillosis in different patient populations. Modified AspICU criteria 2021 and FUNDICU criteria 2024 are applied to non-immunocompromised individuals. The EORTC/MSGERC criteria are used for immunocompromised patients and rely on a combination of host factors, clinical and radiological evidence, and microbiological and histopathological findings to achieve a rapid and accurate diagnosis. For a proven diagnosis of aspergillosis, identification of *Aspergillus* is required through microscopic analysis of sterile material, histopathological or cytopathological examination, or direct microscopic examination of a specimen obtained via needle aspiration or biopsy. Possible aspergillosis requires at least 1 host factor and 1 clinical criterion without any mycological criteria. CSF: Cerebrospinal fluid.

For probable infection, the presence of at least one host factor, one clinical criterion and one mycological criterion is required, while the absence of mycological criteria classified the case as possible infection.

It is important to underline that these guidelines refer to the immunocompromised population and not to the subset of critically ill patients in the ICU who do not fulfil the EORTC/MSG host factors [45].

The population of non-neutropenic, critically ill adult patients is highly heterogenous, including medical and surgical patients, with a wide range of baseline comorbidities and predisposing conditions in ICU (i.e., chronic obstructive pulmonary disease, long-term steroid therapy, hepatic cirrhosis, dialysis, near drowning, or diabetes, sepsis due to bacterial, viral, or parasitic agents, and severe postsepsis immunoparalysis) [68,133]. To address this diagnostic gap, alternative algorithms have been developed. In 2012, the AspICU, and subsequently in 2021, the Modified-AspICU (M-AspICU) algorithm, were proposed to improve IA diagnosis in the ICU settings by substituting the host factors with *Aspergillus*-positive lower respiratory tract specimen culture [7,133,134]. The AspICU criteria consider a combination of clinical signs (such as fever or respiratory deterioration), abnormal imaging, microbiological evidence (including positive culture for *Aspergillus* spp. from lower respiratory tract samples), and host factors not limited to classical immunosuppression [7,133,134].

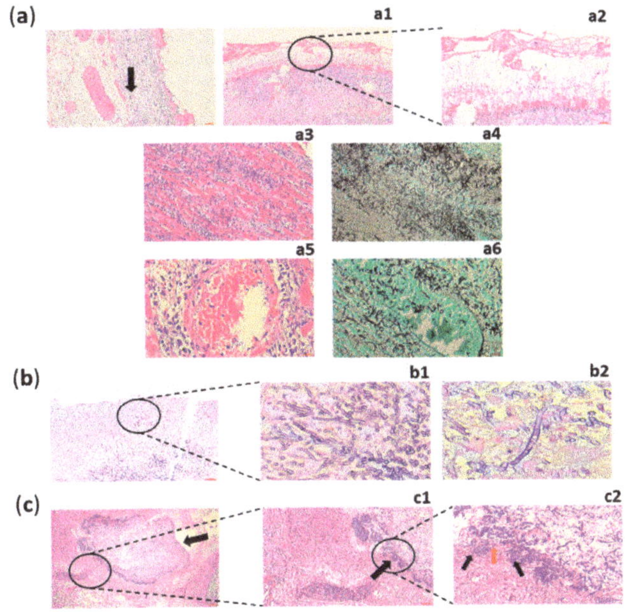

Figure 3. Direct microscopic diagnosis of IA in three immunocompromised patients. (**a**) Patient n. 1 HSCT recipient presenting IA and a granulomatous reaction with giant cells (black arrow) surrounding *Aspergillus* hyphae on the surface of the pleura. (**a1**) Splendor Hoeppli phenomenon (black circle) around the hyphae. This phenomenon consists of radiating eosinophilic material at the lesion's periphery and is typical of chronic granulomatous lesions. (**a2**) Detail of (**a1**): Splendor Hoeppli phenomenon with septated hyphae and conidial heads. (**a3**) Cardiac dissemination of *Aspergillus* hyphae dissecting myocardial fibers with neutrophilic inflammation (myocarditis). (**a4**) Hyphae of *Aspergillus* highlighted by histochemical reaction of Grocott-Gomori methenamine silver (GMS). (**a5**,**a6**): (**a5**) hematogenous spread of *Aspergillus* in IA. (**a6**) Hyphae within a vascular lumen highlighted by GMS. (**b**) Patient n. 2 with AIDS presenting necrotizing pseudomembranous bronchitis with pseudomembrane of *Aspergillus* in bronchial wall eroding the epithelium (black circle; exudate of necrotic tissue, inflammatory cells, fungal hyphae and conidial heads). Figure (**b1**,**b2**) show an enlargement of figure (**b**) of the typical branching at acute angle shape (45°) of the *Aspergillus* hyphae with conidial heads. Hyphae of *Aspergillus* are branching (**b1**) and septated (**b2**). (**c**) Patient n. 3 positive to SARS-CoV-2 with necrotizing pseudomembranous bronchitis. The fungus plug (black arrow) completely occludes the airway lumen. Post-mortem artifacts are present. (**c1**) Enlargement of pseudomembrane of *Aspergillus* growing along the surface of the bronchial mucosa and eroding the epithelium (black arrow; exudate of necrotic tissue, inflammatory cells, fungal hyphae and conidial heads). (**c2**) Detail of (**c1**): fungal hyphae and conidial heads (red arrow) eroding the mucosa (black arrows). Images were taken from the hospital archive.

More recently, to overcome the limitations of AspICU and to further refine the diagnostic classification in ICU settings, the Invasive Fungal Diseases in Adult Patients in ICU (FUNDICU) project was conceived with the aim of developing a standard set of definitions for IFD in non-neutropenic, ICU patients outside the classical immunocompromised patient populations, which could improve the generalizability and comparability of research results [68]. The FUNDICU proposes a more nuanced categorization of fungal infections in ICU patients into three levels: proven, probable, and possible IFD, based on a combination of clinical, radiological, and microbiological findings. Unlike the AspICU algorithm, FUNDICU incorporates advanced diagnostic tools such as galactomannan, β-D-glucan, and PCR from respiratory or blood samples, and recognizes viral coinfections (notably influenza and SARS-CoV-2) as host risk factors for probable IFD [68]. By adapting the diagnostic criteria to the reality of ICU patients, FUNDICU aims to facilitate early recognition of fungal infections and improve the standardization of clinical research endpoints in this high-risk population (Table 2).

Table 2. Comparison of Diagnostic Criteria: EORTC/MSGERC 2020 vs. Modified AspICU 2021 vs. FUNDICU 2024.

	EORTC/MSGERC (2020)	Modified AspICU (2021)	FUNDICU (2024)
Target population	Severely immunocompromised (e.g., neutropenic, HSCT, hematologic malignancies)	Non-neutropenic ICU patients	All critically ill ICU patients, included non-classically immunocompromised
Diagnostic categories	Possible, Probable, Proven	Proven, Probable, Colonization	Possible, Probable, Proven
Required clinical criteria	Not central: diagnosis mostly relies on host factors, imaging and microbiological criteria	Yes: fever, respiratory symptoms, worsening oxygenation	Yes: detailed ICU-adapted criteria (e.g., sepsis, new infiltrates, secretions)
Radiological criteria	Yes: typical signs such as halo, air crescent, cavitation	Yes: infiltrates or new lesions consistent with infection	Yes: including HRCT or other imaging compatible with IPA
Microbiological criteria	Positive culture from sterile site or BAL GM index $\geq$ 0.5 in serum GM index $\geq$ 1.0 in BAL Positive PCR histopathology showing hyphae with tissue damage	BAL/tracheal culture GM in BAL (ODI $\geq$1.0) PCR optional	Culture GM (BAL ODI $\geq$ 1.0, serum ODI $\geq$ 0.5) BDG (>80 pg/mL) PCR on blood or BAL histopathology if available
Included host factors	Strict: Neutropenia (ANC < 500 cells/µL for >10 days), allogeneic HSCT, prolonged corticosteroids, T-cell immunosuppressants, inherited immunodeficiencies, solid organ transplant, recent chemotherapy, GVHD, or treatment with anti-cytokine biologics	ICU-specific: prolonged mechanical ventilation, ARDS, chronic lung disease	Broad: includes viral pneumonia, immunotherapy, ARDS, prolonged ICU stay, chronic lung disease
Advanced diagnostic techniques (PCR, BDG, GM)	Yes (GM, BDG, PCR, histopathology)	Partially, BAL GM included, PCR optional	Yes, required for Probable classification
Recognition of viral co-infections	No	Partially: influenza included, COVID-19 not systematically	Yes: COVID-19 and influenza recognized
Validation and intended use	Standard in trials and guidelines for immunocompromised patients	Used in ICU research, CAPA/IAPA definitions	Built by international consensus, intended for ICU clinical research and standardization

The gold standard for diagnosing aspergillosis is a sterile sample biopsy obtained through bronchoscopy, surgical procedure, or trans-thoracic needle. These procedures can have some executive limitations due to hemodynamic instability, coagulation disorders, or respiratory failure of these patients [124,133,134].

For this reason, the diagnosis should start with the clinical evaluation of patients (host factors, symptoms, and signs of presentation), followed by the execution of diagnostic tests to confirm or not the probability of having aspergillosis.

Usually, clinical presentation involves the pulmonary system with cough, respiratory distress, and dyspnea. Fever and resistance to antibiotics is another sign that is not specific. Chest pain or hemoptysis are frequent in case of angio-invasive growth, which can promote the dissemination to other organs: CNS with convulsions, meningitis, and other clinical manifestations with a bad prognosis [135]. A new radiologic finding or persistent chest infiltrates despite antibiotics should raise suspicion of a superimposed fungal infection.

The role of imaging is to detect typical features to make an early diagnosis of IPA, evaluate the progression, and detect complications. Computed tomography (CT) is the preferred method of diagnosis and follow-up. A non-contrast CT was performed initially, but the contrast is necessary to detect the angio-invasion in the lesion. Radiologic presentation varies about the immune status of patients. Specifically, pulmonary nodules can be observed in both neutropenic and non-neutropenic patients, with a distribution reflecting the infection's behavior (airway invasive or angioinvasive). These nodules may be associated with a halo sign, which, though not specific, serves as an early radiological indicator, particularly in the presence of host risk factors such as neutropenia. The halo sign has a high positive predictive value for diagnosing IPA and is associated with a favorable prognosis. The cavitation or air crescent sign appears later in the infection and is indicative of immune recovery. Other radiologic features linked to pulmonary nodules include hypodense signs and transfissural extension. Consolidation, with or without the reverse halo (atoll) sign or vascular occlusion sign, is another pattern observed in neutropenic patients, though it is not exclusively associated with aspergillosis [136–138].

GM is a polysaccharide in the cell wall of *Aspergillus* spp., incorporated into the wall during fungal growth and released into the circulation when fungus invades the endothelial compartment. There are some differences between the release of GM in neutropenic hosts versus glucocorticoid-treated hosts. In the former, the infection shows an extensive angioinvasion and a high fungal burden; in the latter, the opposite has a significant inflammation role. This evidence explains the differences in the circulating GM amount in these two groups: high in neutropenic and low in steroid-suppressed patients. Platelia *Aspergillus* sandwich enzyme immunoassay is used to detect GM: a monoclonal antibody derived from rats binds to the GM molecule, so the presence of the antigen in the specimen produces a monoclonal antibody-GM-monoclonal antibody complex and the positivity of the test is revealed by the addition of a chromogenic substrate. The value is expressed as ODI.

A positive serum result should be based on a cut-off GM index ≥ 0.5 after testing two separate serum samples or a single sample with a value of ≥ 1.0 [139].

A positive result on BAL fluid should be based on a cut-off GM index ≥ 1.0 after testing two aliquots of a single BAL fluid sample [140]. GM testing sensitivity and specificity depend on the cut-off level used, patient population, and immune status of the host; in particular, its sensitivity in neutropenic patients is higher than in ICU patients without traditional risk factors. Anti-mold prophylaxis/treatment can reduce the sensitivity of the GM assay. It has a prognostic role with a reduction in its value from 7 to 14 days. False negatives are possible in patients treated with antifungal therapy. At the same time, false positives are feasible in patients with another fungal infection. It can be examined on plasma or BAL, and it demonstrates a relatively high negative predictive value (>90%) with a relatively low positive predictive value (<50%) [140].

Results have shown that patients without clinical signs have tested positive for GM, including those who were taking piperacillin/tazobactam, beta-lactams, and other antibiotics. Additionally, patients who were receiving total parenteral nutrition or crystalloid solutions also had false-positive GM results. Furthermore, patients with mucositis or an altered intestinal barrier displayed false-positive GM results due to dietary factors or glucose-containing solutions [141].

1,3-BDG is a cell wall component; it is not specific for *Aspergillus* but is released during IFI. There are many assays, but the guidelines suggest the use of Fungitell®. This assay uses a modified pathway in the *Limulus* Amebocyte lysate (LAL) test by removing factor C from LAL; in the absence of factor C, the coagulation cascade is activated only by the presence of 1,3-BDG; the activity can be measured with a colorimetric method. The positivity is established by a cut-off of ≥80 ng/L detected in at least two consecutive serum samples. Another assay, such as Dynamiker® Fungus (1-3)-β-D-glucan assay (DFA), uses a cut-off lower than Fungitell, 20 pg/mL. Many studies investigated how accurately BDG tests can detect IFI in ICU patients. The results show that there is a lot of variation in how sensitive and specific the tests are. In ICU patients, false positive results can occur due to various clinical factors and conditions such as the use of surgical gauzes, renal replacement therapy, albumin transfusion, and broad-spectrum antibiotics. The specificity and PPV of the test can be improved by conducting 2 consecutive BDG tests, without significantly impacting the NPV. Additionally, combining BDG testing with other fungal biomarkers, such as *Candida albicans* germ tube antibody (CAGTA), or using clinical prediction rules, like the *Candida* score, may enhance the test's effectiveness. As the NPV is generally high, a negative BDG result can be used to stop or not start empirical antifungal therapy. However, the role of a positive BDG result in prompting pre-emptive antifungal strategies is less clear because of the relatively low PPV [142].

False positives are frequent in patients receiving albumin, hemodialysis with cellulose membranes, intravenous immunoglobulin and intravenous amoxicillin-clavulanic acid.

PCR-based methods could be applied to any specimen, showing high sensitivity and specificity for detecting *Aspergillus* DNA. There are several assays with a comparable level of sensibility and specificity. Sensitivity is lower in serum samples than in respiratory specimens, and performance can be reduced in patients with antifungal prevention or treatment. They are also different in the type of population examined. The MycAssay *Aspergillus*® and the AsperGenius® assay are recommended for PCR detection of *Aspergillus* spp. DNA in respiratory samples. These assays are helpful in the case of antifungal resistance in *Aspergillus* spp. They show a sensitivity of almost 78% and a specificity 100% [143].

A novel lateral flow device using an *Aspergillus*-specific monoclonal antibody represents a point-of-care diagnosis of IA. It is quick and cheap. It detects an extracellular glycoprotein secreted by *Aspergillus* spp., during active growth. Serum and BAL are the samples, but it is possible to have a cross-reactivity with *Penicillium*. The main diagnostic tests for aspergillosis, excluding culture-based methods, are summarized in Table 3.

Novel biomarkers or targets based on IA pathogenesis have been studied to discriminate *Aspergillus* infection from those without disease. The rationale is the expression of immune receptors by immune cells involved in the first line of defense against inhaled fungal spores. Cytokine profiles in BAL samples from IPA patients were compared to the control, and the best cytokine was IL-8 with elevated sensitivity (90%), specificity (73%) and NPV (88%) [144].

Table 3. Non-culture test characteristics.

GM in Serum Samples	GM in BAL Samples	1,3-β-D-Glucan	Molecular Methods (PCR)	Lateral Flow Device (LFD) Assay
Best performance in neutropenic patients (ODI > 0.5 in two samples)	Useful in non-neutropenic patients (no positive serum GM)	Nonspecific marker	Detection of several *Aspergillus* spp. in immunocompromised individuals	*Aspergillus*-specific monoclonal antibody for the detection of an extracellular glycoprotein secreted by *Aspergillus* spp., during active growth
Screening test in patient at risk	Cut-off not established: 0.5 USA versus 1.0 Europe in relation to patients' risk	High NPV, useful as screening in high-risk patients	Early diagnosis with a high NPV, high sensitivity	Serum and BAL are the samples
Serum GM present 5–8 days before clinical manifestations	Performance test depends on technique used for BAL procedure (sterile saline volume instilled during bronchoscopy, volume and type of collected BAL fluid), mold prophylaxis or therapy, risk of fungal colonization		It allows to quantify and to recognize *Aspergillus* species	Cross-reactivity with *Penicillium*
Serum GM should not be used on patients at risk but on mold active prophylaxis	Major sensitivity than serum GM assay with a high PPV		High cost and technical difficulties	Rapid and cost-effective

Another biomarker is triacetyl-fusarinine C (TAFC), a fungal molecule produced by some molds. One of these is *A. fumigatus*. It is a secreted siderophore that can be used as a biomarker. In fact, its detection in bodily fluids (e.g., serum, BAL or urine) may aid the diagnosis of IA and represents a promising tool to support the early identification of IPA in high-risk patients; if it has been combined with GM-BAL, the sensitivity and specificity of GM increased [145], but other studies have shown it has no benefit for IPA diagnosis, so its role is controversial. However, TAFC can be measured in urine samples with mass spectrometry.

In case of suspected fungal pneumonia, *Aspergillus* volatile organic compounds, secondary metabolites of varied chemical nature, can be detected in breath: α-trans-bergamotene, β-trans-bergamotene, β-vatirenene-like sesquiterpene or trans-geranyl acetone identified in IPA patients with a sensitivity and specificity of 94% and 93%, respectively [146]. Other studies are necessary to define their roles in diagnostics.

Proteome analysis of BAL shows host and fungal proteins expressed during IPA, which in the future can represent novel biomarkers in *Aspergillus* diagnosis [147].

6. Antifungal Prophylaxis and Pre-Emptive Treatment Versus Antifungal Therapy

The rising number of immunocompromised individuals has led to an increased risk of IFIs, which are associated with a high mortality rate due to various complications. The development of novel chemotherapy agents and immunosuppressive drugs has improved outcomes but also increased susceptibility to IFIs [148].

All of these factors must be taken into account when assessing a high-risk patient, in order to establish a treatment plan that selects the appropriate therapeutic regimen. This includes choosing the right drug based on its potential drug-drug interactions, efficacy, tolerability, and toxicity, while taking into account emerging pharmacological resistance and the introduction of drugs targeting novel molecular and immune pathways [10,149].

In certain cases, antifungal prophylaxis may be considered for patients without signs or symptoms of fungal disease. It is essential to balance the risk of drug-induced side effects and resistance with the potential benefits in terms of increased survival rates. Pre-emptive treatment involves administering antifungal therapy to a population based on laboratory tests that indicate early signs of IFD. Empiric treatment, on the other hand, is administered to patients at risk of fungal infection who present with fever as a sign or symptom of IFD [10].

Different antifungal treatments are available against *Aspergillus* spp. Polyenes such as amphotericin B deoxycholate (AmB) were the first agent substituted by triazoles for their superiority and less toxicity [150]. Initially, voriconazole proved to be superior to amphotericin B in the treatment of IA, also showing fewer side effects [150]. Subsequently, among azoles, isavuconazole demonstrated non-inferiority compared to voriconazole [151], with fewer adverse events, a better pharmacokinetic profile. Finally, posaconazole also showed non-inferiority to voriconazole [152]. Moreover, lipid formulations of AmB, which are less nephrotoxic, are an alternative when voriconazole and isavuconazole are not tolerated or contraindicated. Voriconazole and isavuconazole are available in oral (po) and intravenous (IV) formulations. They represent the primary regiments in IA therapy. A loading dose is necessary for voriconazole: 6 mg/kg IV every 12 h for two doses, followed by 4 mg/kg every 12 h. Therapeutic drug monitoring (TDM) is recommended due to its complex metabolism and numerous drug—drug interactions (DDIs), primarily due to its inhibition of CYP450 enzymes. These interactions can significantly alter the plasma concentrations of both voriconazole and co-administered drugs, potentially leading to important clinical consequences [10]. On the other hand, isavuconazole has a more predictable pharmacokinetic profile and a lower potential for DDIs compared to other azoles. Although it is a moderate inhibitor of CYP3A4, its interaction profile is generally more favorable, making it a safer option in patients receiving multiple concomitant medications. Nonetheless, caution is still required when isavuconazole is co-administered with strong CYP3A4 inhibitors or inducers, as these can affect its plasma levels. Hence, TDM is not always required except in specific situations. For isavuconazole, a loading dose is also needed (372 mg po/IV q8h $\times$ 6 doses), followed by 372 mg po/IV daily [153]. This antifungal agent helps decrease susceptibility to other triazoles, as it shows fewer adverse effects and DDI than voriconazole [151]. Posaconazole is a triazole studied for IFI prophylaxis in high-risk patients and as salvage therapy in refractory fungal infections [154]. Itraconazole is used in non-invasive or chronic forms of aspergillosis or following intolerance or toxicity to other triazoles.

Echinocandins such as caspofungin, micafungin and anidulafungin, available in intravenous formulation, are not recommended as first-line agents but as salvage therapy or in combination with other drugs [155,156].

Combination therapy is a strategy to improve patient outcomes and overcome antifungal resistance, which is often caused by mutations in the target site of triazoles (CYP51A). These mutations can occur due to prolonged therapy or in regions where azoles are used on agricultural products [154,155]. In the presence of azole-resistant infection, lipid AmB with azole or echinocandin have been proposed [156].

In the treatment of aspergillosis in immunocompromised patients, we should consider the different antifungal treatment approaches, evaluate the efficacy of trials of specific drugs in this particular setting, and consider the pharmacological aspect of the available drugs to minimize pharmacological exposure and limit adverse effects [157].

7. Antifungal Resistance

7.1. Amphotericin B

Amphotericin B exerts its antifungal activity by binding to sterols, particularly ergosterol, a key component of the fungal cell membrane. This interaction disrupts membrane integrity through the formation of transmembrane pores, leading to increased permeability and the consequent leakage of intracellular ions and cytoplasmic contents. This membrane dysfunction ultimately results in fungal cell death. Additionally, amphotericin B may induce oxidative stress by generating reactive oxygen species (ROS), further contributing to cellular damage [158,159]. Generally, resistance to amphotericin B is quite rare [160].

Among non-*fumigatus Aspergillus* species, resistance to amphotericin B is mainly observed in *A. terreus* known to exhibit intrinsic resistance, although the underlying molecular mechanisms remain unclear and *A. flavus*. It has been hypothesized that the increased use of amphotericin B, partly driven by the emergence of azole-resistant *A. fumigatus*, may have contributed to the selection and rise of amphotericin B resistance isolates in recent years [158].

7.2. Azoles

Since the late 2000s, azole resistance in *A. fumigatus* has emerged as a serious global concern due to increasing reports of therapeutic failure. Although resistance can develop during antifungal treatment, particularly in patients receiving long-term therapy, environmental exposure to agricultural azole fungicides is considered the primary driver of resistance selection [149,161–163].

Azoles act by inhibiting lanosterol 14α-demethylase, an enzyme encoded by the cyp51A gene, which is essential for ergosterol biosynthesis. Inhibition leads to the accumulation of toxic 14α-methyl sterols, compromising membrane structure and function, and ultimately impairing fungal growth [149,161,162].

Mutations in cyp51A—especially nonsynonymous point mutations—are strongly associated with azole resistance. These mutations can reduce drug binding or uptake, conferring variable resistance profiles [149,161,162].

Due to the natural process of asexual sporulation, *A. fumigatus* populations can harbor diverse genetic variants, including both azole-susceptible and -resistant strains within the same host. This genetic heterogeneity complicates treatment and contributes to the persistence of resistant clones [149,161,162].

7.3. Echinocandins

Echinocandins inhibit β-(1,3)-D-glucan synthase, an enzyme involved in fungal cell wall synthesis. Against *Aspergillus* spp., echinocandins exhibit only fungistatic activity, which limits their utility as monotherapy. They are primarily employed in combination with azoles or polyenes to achieve a synergistic antifungal effect. Resistance to echinocandins among *Aspergillus* spp. remains rare but should still be monitored, especially in cases of prolonged exposure or combination therapy [100,104].

8. Conclusions

In recent years, our understanding of pulmonary aspergillosis in immunocompromised patients has significantly evolved. This narrative review highlights the emergence of new host populations at risk—including non-neutropenic ICU patients and individuals with viral infections such as COVID-19 and influenza—who present with atypical forms of invasive aspergillosis. Diagnostic approaches have progressed beyond traditional culture-based methods, combining different tools such as galactomannan, 1,3-β-D-glucan, PCR, and lateral flow assays. Nevertheless, significant diagnostic delays persist, especially in non-classical hosts, due to limitations in specificity and accessibility of these tools in critically ill patients.

Currently, treatment options for invasive fungal diseases (IFDs) remain limited, and resistance to existing antifungal agents is increasingly common among pathogenic fungi. In particular, the emergence of azole-resistant *A. fumigatus* represents a significant public health concern in Europe [158]. Globally, it is estimated that up to 19% of *A. fumigatus* infections are resistant to azole therapy [165].

The growing threat of azole resistance in *A. fumigatus* is mirrored by a broader rise in antifungal resistance across non-*fumigatus Aspergillus* species and cryptic isolates that can also be intrinsically resistant to antifungal therapies. These developments highlight the urgent need for advanced molecular diagnostics and routine susceptibility testing to ensure timely and appropriate therapeutic decisions.

A new promising drug, Olorofim, has been developed. It is the first antifungal agent in the orotomide class and has demonstrated activity against fungi resistant to approved treatments. It acts by disrupting fungal pyrimidine biosynthesis, ultimately causing cell death. Preliminary data showed the efficacy and the safety of Olorofim for the treatment of IFD in patients with few or no treatment options [166]. Advances in antifungal pharmacology includes also newer triazoles and combination therapies that are reshaping treatment strategies but require further validation in high-risk populations.

Ultimately, improving outcomes for immunocompromised patients with aspergillosis demands a multidisciplinary approach: integrating early risk stratification, timely use of molecular diagnostics, and personalized antifungal treatment. Future research should prioritize prospective studies in emerging risk groups, standardized diagnostic criteria for ICU populations, and strategies to mitigate antifungal resistance globally.

Author Contributions: M.G.B.; writing—review and editing, supervision, L.C.; writing—original draft preparation, G.C.; investigation, writing—review and editing, A.L.; investigation, writing—review and editing, V.S.; investigation, D.R.; investigation, G.V.S.; investigation, G.G.; investigation, S.C. (Stefania Carrara); investigation, A.V.; investigation, C.G.; investigation, F.D.N.; resources and data curation, D.C.; resources and data curation, L.F.; investigation, E.C.; investigation, S.C. (Stefania Cicalini); supervision, C.F.; supervision. All authors have read and agreed to the published version of the manuscript.

Funding: This research received no external funding.

Institutional Review Board Statement: Not applicable.

Informed Consent Statement: Not applicable.

Data Availability Statement: No new data were created or analyzed in this study. Data sharing is not applicable to this article.

Acknowledgments: This research was supported by grants from the Italian Ministry of Health through 'Ricerca Corrente Linea 3'. During the preparation of this manuscript, the author(s) used ChatGPT v.5(OpenAI, San Francisco, CA, USA; https://chat.openai.com (accessed on 10 July 2025)) to assist in editing the manuscript. The authors have reviewed and edited the output and take full responsibility for the content of this publication.

Conflicts of Interest: The authors declare no conflicts of interest.

References

1. Lamoth, F. Aspergillus Fumigatus-Related Species in Clinical Practice. *Front. Microbiol.* **2016**, *7*, 683. [CrossRef]
2. Shi, C.; Shan, Q.; Xia, J.; Wang, L.; Wang, L.; Qiu, L.; Xie, Y.; Lin, N.; Wang, L. Incidence, Risk Factors and Mortality of Invasive Pulmonary Aspergillosis in Patients with Influenza: A Systematic Review and Meta-analysis. *Mycoses* **2022**, *65*, 152–163. [CrossRef]
3. Calderón-Parra, J.; Mills-Sanchez, P.; Moreno-Torres, V.; Tejado-Bravo, S.; Romero-Sánchez, I.; Balandin-Moreno, B.; Calvo-Salvador, M.; Portero-Azorín, F.; García-Masedo, S.; Muñez-Rubio, E.; et al. COVID-19-associated Pulmonary Aspergillosis (CAPA): Risk Factors and Development of a Predictive Score for Critically Ill COVID-19 Patients. *Mycoses* **2022**, *65*, 541–550. [CrossRef]
4. Latgé, J.-P.; Chamilos, G. Aspergillus Fumigatus and Aspergillosis in 2019. *Clin. Microbiol. Rev.* **2019**, *33*, e00140-18. [CrossRef] [PubMed]
5. Geremia, N.; Giovagnorio, F.; Colpani, A.; De Vito, A.; Caruana, G.; Meloni, M.C.; Madeddu, G.; Panese, S.; Parisi, S.G. What Do We Know about Cryptic Aspergillosis? *Microorganisms* **2024**, *12*, 886. [CrossRef] [PubMed]
6. Siopi, M.; Karakatsanis, S.; Roumpakis, C.; Korantanis, K.; Sambatakou, H.; Sipsas, N.V.; Tsirigotis, P.; Pagoni, M.; Meletiadis, J. A Prospective Multicenter Cohort Surveillance Study of Invasive Aspergillosis in Patients with Hematologic Malignancies in Greece: Impact of the Revised EORTC/MSGERC 2020 Criteria. *J. Fungi* **2021**, *7*, 27. [CrossRef]
7. Donnelly, J.P.; Chen, S.C.; Kauffman, C.A.; Steinbach, W.J.; Baddley, J.W.; Verweij, P.E.; Clancy, C.J.; Wingard, J.R.; Lockhart, S.R.; Groll, A.H.; et al. Revision and Update of the Consensus Definitions of Invasive Fungal Disease From the European Organization for Research and Treatment of Cancer and the Mycoses Study Group Education and Research Consortium. *Clin. Infect. Dis.* **2020**, *71*, 1367–1376. [CrossRef]
8. Malesu, V.K. Aspergillosis: A Growing Global Crisis. Available online: https://www.news-medical.net/health/Aspergillosis-A-Growing-Global-Crisis.aspx (accessed on 19 June 2025).
9. Koehler, P.; Bassetti, M.; Chakrabarti, A.; Chen, S.C.A.; Colombo, A.L.; Hoenigl, M.; Klimko, N.; Lass-Flörl, C.; Oladele, R.O.; Vinh, D.C.; et al. Defining and Managing COVID-19-Associated Pulmonary Aspergillosis: The 2020 ECMM/ISHAM Consensus Criteria for Research and Clinical Guidance. *Lancet Infect. Dis.* **2021**, *21*, e149–e162. [CrossRef]
10. Stemler, J.; de Jonge, N.; Skoetz, N.; Sinkó, J.; Brüggemann, R.J.; Busca, A.; Ben-Ami, R.; Ráčil, Z.; Piechotta, V.; Lewis, R.; et al. Antifungal Prophylaxis in Adult Patients with Acute Myeloid Leukaemia Treated with Novel Targeted Therapies: A Systematic Review and Expert Consensus Recommendation from the European Hematology Association. *Lancet Haematol.* **2022**, *9*, e361–e373. [CrossRef] [PubMed]
11. Schauwvlieghe, A.F.A.D.; Rijnders, B.J.A.; Philips, N.; Verwijs, R.; Vanderbeke, L.; Van Tienen, C.; Lagrou, K.; Verweij, P.E.; Van de Veerdonk, F.L.; Gommers, D.; et al. Invasive Aspergillosis in Patients Admitted to the Intensive Care Unit with Severe Influenza: A Retrospective Cohort Study. *Lancet Respir. Med.* **2018**, *6*, 782–792. [CrossRef]
12. Gibbons, J.G.; Rokas, A. The Function and Evolution of the Aspergillus Genome. *Trends Microbiol.* **2013**, *21*, 14–22. [CrossRef]
13. Kousha, M.; Tadi, R.; Soubani, A.O. Pulmonary Aspergillosis: A Clinical Review. *Eur. Respir. Rev.* **2011**, *20*, 156–174. [CrossRef] [PubMed]
14. Latgé, J.P. Aspergillus Fumigatus and Aspergillosis. *Clin. Microbiol. Rev.* **1999**, *12*, 310–350. [CrossRef]
15. Sugui, J.A.; Kwon-Chung, K.J.; Juvvadi, P.R.; Latgé, J.-P.; Steinbach, W.J. Aspergillus Fumigatus and Related Species. *Cold Spring Harb. Perspect. Med.* **2015**, *5*, a019786. [CrossRef]
16. Zhou, S.; Ismail, M.A.I.; Buil, J.B.; Gabr, A.; Verweij, P.E.; Mahgoub, E.-S.; de Hoog, S.; Kang, Y.; Ahmed, S.A. Fungi Involved in Rhinosinusitis in Arid Regions: Insights from Molecular Identification and Antifungal Susceptibility. *Microbiol. Spectr.* **2023**, *11*, e01831-23. [CrossRef]
17. Studies in Mycology No. 59. Available online: https://www.studiesinmycology.org/index.php/issue/61-studies-in-mycology-no-59 (accessed on 15 June 2025).
18. Dagenais, T.R.T.; Keller, N.P. Pathogenesis of Aspergillus Fumigatus in Invasive Aspergillosis. *Clin. Microbiol. Rev.* **2009**, *22*, 447–465. [CrossRef]

19. Tischler, B.Y.; Hohl, T.M. Menacing Mold: Recent Advances in Aspergillus Pathogenesis and Host Defense. *J. Mol. Biol.* **2019**, *431*, 4229–4246. [CrossRef]
20. Erwig, L.P.; Gow, N.A.R. Interactions of Fungal Pathogens with Phagocytes. *Nat. Rev. Microbiol.* **2016**, *14*, 163–176. [CrossRef]
21. Margalit, A.; Kavanagh, K. The Innate Immune Response to Aspergillus Fumigatus at the Alveolar Surface. *FEMS Microbiol. Rev.* **2015**, *39*, 670–687. [CrossRef] [PubMed]
22. Wagener, J.; Wang, X.; Becker, K.L.; Aimanianda, V.; Valsecchi, I.; Gresnigt, M.S.; Netea, M.G.; Latge, J.-P.; Gow, N.A.R.; van de Veerdonk, F.L. Immunomodulatory Function of Chitosan Is Dependent on Complement Receptor 3. *Cell Surf.* **2025**, *14*, 100146. [CrossRef] [PubMed]
23. Garlanda, C.; Bottazzi, B.; Magrini, E.; Inforzato, A.; Mantovani, A. PTX3, a Humoral Pattern Recognition Molecule, in Innate Immunity, Tissue Repair and Cancer. *Physiol. Rev.* **2018**, *98*, 623–639. [CrossRef]
24. Espinosa, V.; Jhingran, A.; Dutta, O.; Kasahara, S.; Donnelly, R.; Du, P.; Rosenfeld, J.; Leiner, I.; Chen, C.-C.; Ron, Y.; et al. Inflammatory Monocytes Orchestrate Innate Antifungal Immunity in the Lung. *PLoS Pathog.* **2014**, *10*, e1003940. [CrossRef] [PubMed]
25. Jaillon, S.; Peri, G.; Delneste, Y.; Frémaux, I.; Doni, A.; Moalli, F.; Garlanda, C.; Romani, L.; Gascan, H.; Bellocchio, S.; et al. The Humoral Pattern Recognition Receptor PTX3 Is Stored in Neutrophil Granules and Localizes in Extracellular Traps. *J. Exp. Med.* **2007**, *204*, 793–804. [CrossRef] [PubMed]
26. Urban, C.F.; Nett, J.E. Neutrophil Extracellular Traps in Fungal Infection. *Semin. Cell Dev. Biol.* **2019**, *89*, 47–57. [CrossRef]
27. Bozza, S.; Gaziano, R.; Lipford, G.B.; Montagnoli, C.; Bacci, A.; Di Francesco, P.; Kurup, V.P.; Wagner, H.; Romani, L. Vaccination of Mice against Invasive Aspergillosis with Recombinant Aspergillus Proteins and CpG Oligodeoxynucleotides as Adjuvants. *Microbes Infect.* **2002**, *4*, 1281–1290. [CrossRef]
28. Hohl, T.M. Immune Responses to Invasive Aspergillosis: New Understanding and Therapeutic Opportunities. *Curr. Opin. Infect. Dis.* **2017**, *30*, 364–371. [CrossRef]
29. Heung, L.J.; Wiesner, D.L.; Wang, K.; Rivera, A.; Hohl, T.M. Immunity to Fungi in the Lung. *Semin. Immunol.* **2023**, *66*, 101728. [CrossRef]
30. Dewi, I.M.W.; van de Veerdonk, F.L.; Gresnigt, M.S. The Multifaceted Role of T-Helper Responses in Host Defense against Aspergillus Fumigatus. *J. Fungi* **2017**, *3*, 55. [CrossRef]
31. Romani, L. Immunity to Fungal Infections. *Nat. Rev. Immunol.* **2011**, *11*, 275–288. [CrossRef] [PubMed]
32. Azoulay, E.; Afessa, B. The Intensive Care Support of Patients with Malignancy: Do Everything That Can Be Done. *Intensive Care Med.* **2006**, *32*, 3–5. [CrossRef]
33. Shariati, A.; Moradabadi, A.; Chegini, Z.; Khoshbayan, A.; Didehdar, M. An Overview of the Management of the Most Important Invasive Fungal Infections in Patients with Blood Malignancies. *Infect. Drug Resist.* **2020**, *13*, 2329–2354. [CrossRef] [PubMed]
34. Bays, D.J.; Thompson, G.R. Fungal Infections of the Stem Cell Transplant Recipient and Hematologic Malignancy Patients. *Infect. Dis. Clin. N. Am.* **2019**, *33*, 545–566. [CrossRef] [PubMed]
35. Azoulay, E.; Mokart, D.; Pène, F.; Lambert, J.; Kouatchet, A.; Mayaux, J.; Vincent, F.; Nyunga, M.; Bruneel, F.; Laisne, L.-M.; et al. Outcomes of Critically Ill Patients with Hematologic Malignancies: Prospective Multicenter Data from France and Belgium—A Groupe de Recherche Respiratoire En Réanimation Onco-Hématologique Study. *J. Clin. Oncol.* **2013**, *31*, 2810–2818. [CrossRef]
36. Blot, S.I.; Taccone, F.S.; Van den Abeele, A.-M.; Bulpa, P.; Meersseman, W.; Brusselaers, N.; Dimopoulos, G.; Paiva, J.A.; Misset, B.; Rello, J.; et al. A Clinical Algorithm to Diagnose Invasive Pulmonary Aspergillosis in Critically Ill Patients. *Am. J. Respir. Crit. Care Med.* **2012**, *186*, 56–64. [CrossRef]
37. Colombo, A.L.; de Almeida Júnior, J.N.; Slavin, M.A.; Chen, S.C.-A.; Sorrell, T.C. Candida and Invasive Mould Diseases in Non-Neutropenic Critically Ill Patients and Patients with Haematological Cancer. *Lancet Infect. Dis.* **2017**, *17*, e344–e356. [CrossRef]
38. Ullmann, A.J.; Lipton, J.H.; Vesole, D.H.; Chandrasekar, P.; Langston, A.; Tarantolo, S.R.; Greinix, H.; Morais de Azevedo, W.; Reddy, V.; Boparai, N.; et al. Posaconazole or Fluconazole for Prophylaxis in Severe Graft-versus-Host Disease. *N. Engl. J. Med.* **2007**, *356*, 335–347. [CrossRef] [PubMed]
39. Cornely, O.A.; Maertens, J.; Winston, D.J.; Perfect, J.; Ullmann, A.J.; Walsh, T.J.; Helfgott, D.; Holowiecki, J.; Stockelberg, D.; Goh, Y.-T.; et al. Posaconazole vs. Fluconazole or Itraconazole Prophylaxis in Patients with Neutropenia. *N. Engl. J. Med.* **2007**, *356*, 348–359. [CrossRef]
40. Crawford, J.; Dale, D.C.; Lyman, G.H. Chemotherapy-Induced Neutropenia: Risks, Consequences, and New Directions for Its Management. *Cancer* **2004**, *100*, 228–237. [CrossRef]
41. Newburger, P.E.; Dale, D.C. Evaluation and Management of Patients with Isolated Neutropenia. *Semin. Hematol.* **2013**, *50*, 198–206. [CrossRef]

42. Kontoyiannis, D.P.; [illegible] S.P.; Wierda, W.G.; Wright, S.; Albert, N.D.; Ferrajoli, A.; Keating, M.; Lewis, R.E. Impaired Bactericidal but Not Fungicidal Activity of Polymorphonuclear Neutrophils in Patients with Chronic Lymphocytic Leukemia. *Leuk. Lymphoma* **2013**, *54*, 1730–1733. [CrossRef]
43. Beyer, M.; Kochanek, M.; Darabi, K.; Popov, A.; Jensen, M.; Endl, E.; Knolle, P.A.; Thomas, R.K.; von Bergwelt-Baildon, M.; Debey, S.; et al. Reduced Frequencies and Suppressive Function of CD4+CD25hi Regulatory T Cells in Patients with Chronic Lymphocytic Leukemia after Therapy with Fludarabine. *Blood* **2005**, *106*, 2018–2025. [CrossRef]
44. Neofytos, D.; Lu, K.; Hatfield-Seung, A.; Blackford, A.; Marr, K.A.; Treadway, S.; Ostrander, D.; Nussenblatt, V.; Karp, J. Epidemiology, Outcomes, and Risk Factors of Invasive Fungal Infections in Adult Patients with Acute Myelogenous Leukemia after Induction Chemotherapy. *Diagn. Microbiol. Infect. Dis.* **2013**, *75*, 144–149. [CrossRef]
45. Douglas, A.P.; Smibert, O.C.; Bajel, A.; Halliday, C.L.; Lavee, O.; McMullan, B.; Yong, M.K.; van Hal, S.J.; Chen, S.C.-A. Australasian Antifungal Guidelines Steering Committee Consensus Guidelines for the Diagnosis and Management of Invasive Aspergillosis, 2021. *Intern. Med. J.* **2021**, *51* (Suppl. 7), 143–176. [CrossRef]
46. Lamoth, F.; Calandra, T. Pulmonary Aspergillosis: Diagnosis and Treatment. *Eur. Respir. Rev.* **2022**, *31*, 220114. [CrossRef] [PubMed]
47. Bukkems, L.M.P.; van Dommelen, L.; Regis, M.; van den Heuvel, E.; Nieuwenhuizen, L. The Use of Galactomannan Antigen Assays for the Diagnosis of Invasive Pulmonary Aspergillosis in the Hematological Patient: A Systematic Review and Meta-Analysis. *J. Fungi* **2023**, *9*, 674. [CrossRef] [PubMed]
48. Rüping, M.J.G.T.; Vehreschild, J.J.; Cornely, O.A. Patients at High Risk of Invasive Fungal Infections: When and How to Treat. *Drugs* **2008**, *68*, 1941–1962. [CrossRef]
49. Rajasingham, R.; Smith, R.M.; Park, B.J.; Jarvis, J.N.; Govender, N.P.; Chiller, T.M.; Denning, D.W.; Loyse, A.; Boulware, D.R. Global Burden of Disease of HIV-Associated Cryptococcal Meningitis: An Updated Analysis. *Lancet Infect. Dis.* **2017**, *17*, 873–881. [CrossRef] [PubMed]
50. Colombo, A.L.; Tobón, A.; Restrepo, A.; Queiroz-Telles, F.; Nucci, M. Epidemiology of Endemic Systemic Fungal Infections in Latin America. *Med. Mycol.* **2011**, *49*, 785–798. [CrossRef]
51. Buchacz, K.; Baker, R.K.; Palella, F.J.; Chmiel, J.S.; Lichtenstein, K.A.; Novak, R.M.; Wood, K.C.; Brooks, J.T. HOPS Investigators AIDS-Defining Opportunistic Illnesses in US Patients, 1994–2007: A Cohort Study. *AIDS* **2010**, *24*, 1549–1559. [CrossRef]
52. Antinori, S.; Nebuloni, M.; Magni, C.; Fasan, M.; Adorni, F.; Viola, A.; Corbellino, M.; Galli, M.; Vago, G.; Parravicini, C.; et al. Trends in the Postmortem Diagnosis of Opportunistic Invasive Fungal Infections in Patients with AIDS: A Retrospective Study of 1,630 Autopsies Performed between 1984 and 2002. *Am. J. Clin. Pathol.* **2009**, *132*, 221–227. [CrossRef]
53. Kosmidis, C.; Denning, D.W. The Clinical Spectrum of Pulmonary Aspergillosis. *Thorax* **2015**, *70*, 270–277. [CrossRef]
54. Goncer Rodriguez, I. Aspergillosis in AIDS. Aspergillus and Aspergillosis. Available online: https://www.Aspergillus.org.uk/new_treatment/aspergillosis-in-aids/ (accessed on 17 January 2025).
55. Keating, J.J.; Rogers, T.; Petrou, M.; Cartledge, J.D.; Woodrow, D.; Nelson, M.; Hawkins, D.A.; Gazzard, B.G. Management of Pulmonary Aspergillosis in AIDS: An Emerging Clinical Problem. *J. Clin. Pathol.* **1994**, *47*, 805–809. [CrossRef]
56. Holding, K.J.; Dworkin, M.S.; Wan, P.C.; Hanson, D.L.; Klevens, R.M.; Jones, J.L.; Sullivan, P.S. Aspergillosis among People Infected with Human Immunodeficiency Virus: Incidence and Survival. Adult and Adolescent Spectrum of HIV Disease Project. *Clin. Infect. Dis.* **2000**, *31*, 1253–1257. [CrossRef]
57. Tong, K.B.; Lau, C.J.; Murtagh, K.; Layton, A.J.; Seifeldin, R. The Economic Impact of Aspergillosis: Analysis of Hospital Expenditures across Patient Subgroups. *Int. J. Infect. Dis.* **2009**, *13*, 24–36. [CrossRef]
58. Baddley, J.W. Clinical Risk Factors for Invasive Aspergillosis. *Med. Mycol.* **2011**, *49* (Suppl. 1), S7–S12. [CrossRef] [PubMed]
59. Miller, W.T.; Sais, G.J.; Frank, I.; Gefter, W.B.; Aronchick, J.M.; Miller, W.T. Pulmonary Aspergillosis in Patients with AIDS. Clinical and Radiographic Correlations. *Chest* **1994**, *105*, 37–44. [CrossRef] [PubMed]
60. Barry, S.M.; Lipman, M.C.; Johnson, M.A.; Prentice, H.G. Respiratory Infections in Immunocompromised Patients. *Curr. Opin. Pulm. Med.* **1999**, *5*, 168–173. [CrossRef] [PubMed]
61. Mylonakis, E.; Barlam, T.F.; Flanigan, T.; Rich, J.D. Pulmonary Aspergillosis and Invasive Disease in AIDS: Review of 342 Cases. *Chest* **1998**, *114*, 251–262. [CrossRef]
62. Woitas, R.P.; Rockstroh, J.K.; Theisen, A.; Leutner, C.; Sauerbruch, T.; Spengler, U. Changing Role of Invasive Aspergillosis in AIDS—A Case Control Study. *J. Infect.* **1998**, *37*, 116–122. [CrossRef]
63. Wallace, J.M.; Lim, R.; Browdy, B.L.; Hopewell, P.C.; Glassroth, J.; Rosen, M.J.; Reichman, L.B.; Kvale, P.A. Risk Factors and Outcomes Associated with Identification of Aspergillus in Respiratory Specimens from Persons with HIV Disease. Pulmonary Complications of HIV Infection Study Group. *Chest* **1998**, *114*, 131–137. [CrossRef]
64. Doumbo, S.N.; Cissoko, Y.; Dama, S.; Niangaly, A.; Garango, A.; Konaté, A.; Koné, A.; Traoré, B.; Thera, M.; Djimde, A.; et al. The Estimated Burden of Fungal Diseases in Mali. *J. Mycol. Med.* **2023**, *33*, 101333. [CrossRef]

65. Truda, V.S.S.; Falci, D.R.; Porfírio, F.M.V.; de Santos, D.W.d.C.L.; Junior, F.I.O.; Pasqualotto, A.C.; Puga, F.G.; Bollela, V.R.; Junior, J.N.A.; Ferreira, P.R.A.; et al. A Contemporary Investigation of Burden and Natural History of Aspergillosis in People Living with HIV/AIDS. *Mycoses* **2023**, *66*, 632–638. [CrossRef] [PubMed]
66. Yerbanga, I.W.; Nakanabo Diallo, S.; Rouamba, T.; Denis, O.; Rodriguez-Villalobos, H.; Montesinos, I.; Bamba, S. A Systematic Review of Epidemiology, Risk Factors, Diagnosis, Antifungal Resistance, and Management of Invasive Aspergillosis in Africa. *J. Mycol. Med.* **2023**, *33*, 101328. [CrossRef] [PubMed]
67. Denning, D.W.; Morgan, E.F. Quantifying Deaths from Aspergillosis in HIV Positive People. *J. Fungi* **2022**, *8*, 1131. [CrossRef]
68. Bassetti, M.; Giacobbe, D.R.; Agvald-Ohman, C.; Akova, M.; Alastruey-Izquierdo, A.; Arikan-Akdagli, S.; Azoulay, E.; Blot, S.; Cornely, O.A.; Cuenca-Estrella, M.; et al. Invasive Fungal Diseases in Adult Patients in Intensive Care Unit (FUNDICU): 2024 Consensus Definitions from ESGCIP, EFISG, ESICM, ECMM, MSGERC, ISAC, and ISHAM. *Intensive Care Med.* **2024**, *50*, 502–515. [CrossRef]
69. Loreto, E.S.; Tondolo, J.S.M.; Loreto, E.S.; Tondolo, J.S.M. Introductory Chapter: Epidemiology of Invasive Fungal Infection—An Overview. In *Fungal Infection*; IntechOpen: London, UK, 2019; ISBN 978-1-83880-469-5.
70. Singh, N.; Husain, S. AST Infectious Diseases Community of Practice Aspergillosis in Solid Organ Transplantation. *Am. J. Transpl.* **2013**, *13* (Suppl. 4), 228–241. [CrossRef]
71. Lum, L.; Lee, A.; Vu, M.; Strasser, S.; Davis, R. Epidemiology and Risk Factors for Invasive Fungal Disease in Liver Transplant Recipients in a Tertiary Transplant Center. *Transpl. Infect. Dis.* **2020**, *22*, e13361. [CrossRef]
72. Gioia, F.; Filigheddu, E.; Corbella, L.; Fernández-Ruiz, M.; López-Medrano, F.; Pérez-Ayala, A.; Aguado, J.M.; Fariñas, M.C.; Arnaiz, F.; Calvo, J.; et al. Invasive Aspergillosis in Solid Organ Transplantation: Diagnostic Challenges and Differences in Outcome in a Spanish National Cohort (Diaspersot Study). *Mycoses* **2021**, *64*, 1334–1345. [CrossRef] [PubMed]
73. Abe, K.; Shinoda, M.; Uno, S.; Obara, H.; Kitago, M.; Abe, Y.; Hishida, T.; Yagi, H.; Hasegawa, Y.; Kitagawa, Y. Invasive Pulmonary Aspergillosis after Liver Transplantation: Lessons from Successfully Treated Cases and Review of the Literature. *Surg. Today* **2021**, *51*, 1361–1370. [CrossRef]
74. Singh, N.; Avery, R.K.; Munoz, P.; Pruett, T.L.; Alexander, B.; Jacobs, R.; Tollemar, J.G.; Dominguez, E.A.; Yu, C.M.; Paterson, D.L.; et al. Trends in Risk Profiles for and Mortality Associated with Invasive Aspergillosis among Liver Transplant Recipients. *Clin. Infect. Dis.* **2003**, *36*, 46–52. [CrossRef]
75. Farges, C.; Cointault, O.; Murris, M.; Lavayssiere, L.; Lakhdar-Ghazal, S.; Del Bello, A.; Hebral, A.-L.; Esposito, L.; Nogier, M.-B.; Sallusto, F.; et al. Outcomes of Solid Organ Transplant Recipients with Invasive Aspergillosis and Other Mold Infections. *Transpl. Infect. Dis.* **2020**, *22*, e13200. [CrossRef]
76. Fortún, J.; Martín-Dávila, P.; Moreno, S.; De Vicente, E.; Nuño, J.; Candelas, A.; Bárcena, R.; García, M. Risk Factors for Invasive Aspergillosis in Liver Transplant Recipients. *Liver Transpl.* **2002**, *8*, 1065–1070. [CrossRef]
77. Kusne, S.; Dummer, J.S.; Singh, N.; Makowka, L.; Esquivel, C.; Starzl, T.E.; Ho, M. Fungal Infections After Liver Transplantation. *Transpl. Proc.* **1988**, *20*, 650–651.
78. Neofytos, D.; Chatzis, O.; Nasioudis, D.; Boely Janke, E.; Doco Lecompte, T.; Garzoni, C.; Berger, C.; Cussini, A.; Boggian, K.; Khanna, N.; et al. Epidemiology, Risk Factors and Outcomes of Invasive Aspergillosis in Solid Organ Transplant Recipients in the Swiss Transplant Cohort Study. *Transpl. Infect. Dis.* **2018**, *20*, e12898. [CrossRef] [PubMed]
79. Albillos, A.; Lario, M.; Álvarez-Mon, M. Cirrhosis-Associated Immune Dysfunction: Distinctive Features and Clinical Relevance. *J. Hepatol.* **2014**, *61*, 1385–1396. [CrossRef]
80. de la Garza, R.G.; Sarobe, P.; Merino, J.; Lasarte, J.J.; D'Avola, D.; Belsue, V.; Delgado, J.A.; Silva, L.; Iñarrairaegui, M.; Sangro, B.; et al. Trial of Complete Weaning from Immunosuppression for Liver Transplant Recipients: Factors Predictive of Tolerance. *Liver Transpl.* **2013**, *19*, 937–944. [CrossRef] [PubMed]
81. Reddy, M.S.; Varghese, J.; Venkataraman, J.; Rela, M. Matching Donor to Recipient in Liver Transplantation: Relevance in Clinical Practice. *World J. Hepatol.* **2013**, *5*, 603–611. [CrossRef] [PubMed]
82. Singh, N.; Arnow, P.M.; Bonham, A.; Dominguez, E.; Paterson, D.L.; Pankey, G.A.; Wagener, M.M.; Yu, V.L. Invasive Aspergillosis in Liver Transplant Recipients in the 1990s. *Transplantation* **1997**, *64*, 716–720. [CrossRef]
83. Gayowski, T.; Marino, I.R.; Singh, N.; Doyle, H.; Wagener, M.; Fung, J.J.; Starzl, T.E. Orthotopic Liver Transplantation in High-Risk Patients: Risk Factors Associated with Mortality and Infectious Morbidity. *Transplantation* **1998**, *65*, 499–504. [CrossRef]
84. Saliba, F.; Delvart, V.; Ichaï, P.; Kassis, N.; Botterel, F.; Mihaila, L.; Azoulay, D.; Adam, R.; Castaing, D.; Bretagne, S.; et al. Fungal Infections after Liver Transplantation: Outcomes and Risk Factors Revisited in the MELD Era. *Clin. Transpl.* **2013**, *27*, E454–E461. [CrossRef]
85. Melenotte, C.; Aimanianda, V.; Slavin, M.; Aguado, J.M.; Armstrong-James, D.; Chen, Y.-C.; Husain, S.; Van Delden, C.; Saliba, F.; Lefort, A.; et al. Invasive Aspergillosis in Liver Transplant Recipients. *Transpl. Infect. Dis.* **2023**, *25*, e14049. [CrossRef] [PubMed]

86. Gavalda, J.; Len, O.; San Juan, R.; Aguado, J.M.; Fortún, J.; Lumbreras, C.; Moreno, A.; Munoz, P.; Blanes, M.; Ramos, A.; et al. Risk Factors for Invasive Aspergillosis in Solid-Organ Transplant Recipients: A Case-Control Study. *Clin. Infect. Dis.* **2005**, *41*, 52–59. [CrossRef] [PubMed]
87. Bonham, C.A.; Dominguez, E.A.; Fukui, M.B.; Paterson, D.L.; Pankey, G.A.; Wagener, M.M.; Fung, J.J.; Singh, N. Central Nervous System Lesions in Liver Transplant Recipients: Prospective Assessment of Indications for Biopsy and Implications for Management. *Transplantation* **1998**, *66*, 1596–1604. [CrossRef]
88. Altiparmak, M.R.; Apaydin, S.; Trablus, S.; Serdengecti, K.; Ataman, R.; Ozturk, R.; Erek, E. Systemic Fungal Infections after Renal Transplantation. *Scand. J. Infect. Dis.* **2002**, *34*, 284–288. [CrossRef]
89. Brown, R.S.; Lake, J.R.; Katzman, B.A.; Ascher, N.L.; Somberg, K.A.; Emond, J.C.; Roberts, J.P. Incidence and Significance of *Aspergillus* Cultures Following Liver and Kidney Transplantation. *Transplantation* **1996**, *61*, 666–669. [CrossRef]
90. Neofytos, D.; Garcia-Vidal, C.; Lamoth, F.; Lichtenstern, C.; Perrella, A.; Vehreschild, J.J. Invasive Aspergillosis in Solid Organ Transplant Patients: Diagnosis, Prophylaxis, Treatment, and Assessment of Response. *BMC Infect. Dis.* **2021**, *21*, 296. [CrossRef] [PubMed]
91. Lortholary, O.; Gangneux, J.-P.; Sitbon, K.; Lebeau, B.; de Monbrison, F.; Le Strat, Y.; Coignard, B.; Dromer, F.; Bretagne, S. French Mycosis Study Group Epidemiological Trends in Invasive Aspergillosis in France: The SAIF Network (2005–2007). *Clin. Microbiol. Infect.* **2011**, *17*, 1882–1889. [CrossRef]
92. Fishman, J.A.; Issa, N.C. Infection in Organ Transplantation: Risk Factors and Evolving Patterns of Infection. *Infect. Dis. Clin. N. Am.* **2010**, *24*, 273–283. [CrossRef]
93. Eckardt, K.-U.; Kasiske, B.L.; Zeier, M.G. Kidney Disease: Improving Global Outcomes (KDIGO) Transplant Work Group KDIGO Clinical Practice Guideline for the Care of Kidney Transplant Recipients. *Am. J. Transpl.* **2009**, *9* (Suppl. 3), S1–S155. [CrossRef]
94. Ojo, A.O.; Hanson, J.A.; Wolfe, R.A.; Leichtman, A.B.; Agodoa, L.Y.; Port, F.K. Long-Term Survival in Renal Transplant Recipients with Graft Function. *Kidney Int.* **2000**, *57*, 307–313. [CrossRef]
95. López-Medrano, F.; Fernández-Ruiz, M.; Silva, J.T.; Carver, P.L.; van Delden, C.; Merino, E.; Pérez-Saez, M.J.; Montero, M.; Coussement, J.; de Abreu Mazzolin, M.; et al. Multinational Case-Control Study of Risk Factors for the Development of Late Invasive Pulmonary Aspergillosis Following Kidney Transplantation. *Clin. Microbiol. Infect.* **2018**, *24*, 192–198. [CrossRef]
96. Panackal, A.A.; Dahlman, A.; Keil, K.T.; Peterson, C.L.; Mascola, L.; Mirza, S.; Phelan, M.; Lasker, B.A.; Brandt, M.E.; Carpenter, J.; et al. Outbreak of Invasive Aspergillosis among Renal Transplant Recipients. *Transplantation* **2003**, *75*, 1050–1053. [CrossRef] [PubMed]
97. Husain, S.; Paterson, D.L.; Studer, S.; Pilewski, J.; Crespo, M.; Zaldonis, D.; Shutt, K.; Pakstis, D.L.; Zeevi, A.; Johnson, B.; et al. Voriconazole Prophylaxis in Lung Transplant Recipients. *Am. J. Transpl.* **2006**, *6*, 3008–3016. [CrossRef]
98. Solé, A.; Morant, P.; Salavert, M.; Pemán, J.; Morales, P. Valencia Lung Transplant Group *Aspergillus* Infections in Lung Transplant Recipients: Risk Factors and Outcome. *Clin. Microbiol. Infect.* **2005**, *11*, 359–365. [CrossRef]
99. Aguilar, C.A.; Hamandi, B.; Fegbeutel, C.; Silveira, F.P.; Verschuuren, E.A.; Ussetti, P.; Chin-Hong, P.V.; Sole, A.; Holmes-Liew, C.; Billaud, E.M.; et al. Clinical Risk Factors for Invasive Aspergillosis in Lung Transplant Recipients: Results of an International Cohort Study. *J. Heart Lung Transpl.* **2018**, *37*, 1226–1234. [CrossRef] [PubMed]
100. Krenke, R.; Grabczak, E.M. Tracheobronchial Manifestations of Aspergillus Infections. *Sci. World J.* **2011**, *11*, 2310–2329. [CrossRef] [PubMed]
101. Shields, R.K.; Nguyen, M.H.; Shullo, M.A.; Silveira, F.P.; Kwak, E.J.; Abdel Massih, R.C.; Toyoda, Y.; Bermudez, C.A.; Bhama, J.K.; Kormos, R.L.; et al. Invasive Aspergillosis among Heart Transplant Recipients Is Rare but Causes Rapid Death Due to Septic Shock and Multiple Organ Dysfunction Syndrome. *Scand. J. Infect. Dis.* **2012**, *44*, 982–986. [CrossRef]
102. Muñoz, P.; Cerón, I.; Valerio, M.; Palomo, J.; Villa, A.; Eworo, A.; Fernández-Yáñez, J.; Guinea, J.; Bouza, E. Invasive Aspergillosis among Heart Transplant Recipients: A 24-Year Perspective. *J. Heart Lung Transpl.* **2014**, *33*, 278–288. [CrossRef]
103. Montoya, J.G.; Chaparro, S.V.; Celis, D.; Cortés, J.A.; Leung, A.N.; Robbins, R.C.; Stevens, D.A. Invasive Aspergillosis in the Setting of Cardiac Transplantation. *Clin. Infect. Dis.* **2003**, *37* (Suppl. 3), S281–S292. [CrossRef]
104. Muñoz, P.; Singh, N.; Bouza, E. Treatment of Solid Organ Transplant Patients with Invasive Fungal Infections: Should a Combination of Antifungal Drugs Be Used? *Curr. Opin. Infect. Dis.* **2006**, *19*, 365–370. [CrossRef]
105. Lu, L.Y.; Lee, H.M.; Burke, A.; Li Bassi, G.; Torres, A.; Fraser, J.F.; Fanning, J.P. Prevalence, Risk Factors, Clinical Features, and Outcome of Influenza-Associated Pulmonary Aspergillosis in Critically Ill Patients: A Systematic Review and Meta-Analysis. *Chest* **2024**, *165*, 540–558. [CrossRef] [PubMed]
106. COVID-19 Map. Johns Hopkins Coronavirus Resource Center. Available online: https://coronavirus.jhu.edu/map.html. (accessed on 21 January 2025).
107. Costela-Ruiz, V.J.; Illescas-Montes, R.; Puerta-Puerta, J.M.; Ruiz, C.; Melguizo-Rodríguez, L. SARS-CoV-2 Infection: The Role of Cytokines in COVID-19 Disease. *Cytokine Growth Factor Rev.* **2020**, *54*, 62–75. [CrossRef]

108. Gil-Etayo, F.J.; Suàrez-Fernández, P.; Cabrera-Marante, O.; Arroyo, D.; Garcinuño, S.; Naranjo, L.; Pleguezuelo, D.E.; Allende, L.M.; Mancebo, E.; Lalueza, A.; et al. T-Helper Cell Subset Response Is a Determining Factor in COVID-19 Progression. *Front. Cell Infect. Microbiol.* **2021**, *11*, 624483. [CrossRef]
109. Chen, G.; Wu, D.; Guo, W.; Cao, Y.; Huang, D.; Wang, H.; Wang, T.; Zhang, X.; Chen, H.; Yu, H.; et al. Clinical and Immunological Features of Severe and Moderate Coronavirus Disease 2019. *J. Clin. Investig.* **2020**, *130*, 2620–2629. [CrossRef]
110. Bartoletti, M.; Pascale, R.; Cricca, M.; Rinaldi, M.; Maccaro, A.; Bussini, L.; Fornaro, G.; Tonetti, T.; Pizzilli, G.; Francalanci, E.; et al. Epidemiology of Invasive Pulmonary Aspergillosis Among Intubated Patients with COVID-19: A Prospective Study. *Clin. Infect. Dis.* **2021**, *73*, e3606–e3614. [CrossRef]
111. White, P.L.; Dhillon, R.; Cordey, A.; Hughes, H.; Faggian, F.; Soni, S.; Pandey, M.; Whitaker, H.; May, A.; Morgan, M.; et al. A National Strategy to Diagnose Coronavirus Disease 2019-Associated Invasive Fungal Disease in the Intensive Care Unit. *Clin. Infect. Dis.* **2021**, *73*, e1634–e1644. [CrossRef]
112. Montrucchio, G.; Lupia, T.; Lombardo, D.; Stroffolini, G.; Corcione, S.; De Rosa, F.G.; Brazzi, L. Risk Factors for Invasive Aspergillosis in ICU Patients with COVID-19: Current Insights and New Key Elements. *Ann. Intensive Care* **2021**, *11*, 136. [CrossRef]
113. Pons, S.; Fodil, S.; Azoulay, E.; Zafrani, L. The Vascular Endothelium: The Cornerstone of Organ Dysfunction in Severe SARS-CoV-2 Infection. *Crit. Care* **2020**, *24*, 353. [CrossRef]
114. Lee, J.S.; Park, S.; Jeong, H.W.; Ahn, J.Y.; Choi, S.J.; Lee, H.; Choi, B.; Nam, S.K.; Sa, M.; Kwon, J.-S.; et al. Immunophenotyping of COVID-19 and Influenza Highlights the Role of Type I Interferons in Development of Severe COVID-19. *Sci. Immunol.* **2020**, *5*, eabd1554. [CrossRef] [PubMed]
115. Chavda, V.P.; Mishra, T.; Kamaraj, S.; Punetha, S.; Sengupta, O.; Joshi, Y.; Vuppu, S.; Vaghela, D.; Vora, L. Post-COVID-19 Fungal Infection in the Aged Population. *Vaccines* **2023**, *11*, 555. [CrossRef] [PubMed]
116. Hughes, S.; Troise, O.; Donaldson, H.; Mughal, N.; Moore, L.S.P. Bacterial and Fungal Coinfection among Hospitalized Patients with COVID-19: A Retrospective Cohort Study in a UK Secondary-Care Setting. *Clin. Microbiol. Infect.* **2020**, *26*, 1395–1399. [CrossRef]
117. Garcia-Vidal, C.; Sanjuan, G.; Moreno-García, E.; Puerta-Alcalde, P.; Garcia-Pouton, N.; Chumbita, M.; Fernandez-Pittol, M.; Pitart, C.; Inciarte, A.; Bodro, M.; et al. Incidence of Co-Infections and Superinfections in Hospitalized Patients with COVID-19: A Retrospective Cohort Study. *Clin. Microbiol. Infect.* **2021**, *27*, 83–88. [CrossRef]
118. Paul, M.; Sasidharan, J.; Taneja, J.; Chatterjee, K.; Abbas, S.Z.; Chowdhury, V.; Das, A. Invasive Mucormycosis and Aspergillosis Coinfection Associated with Post-COVID-19 Pneumonia in a Tertiary Care Hospital. *Med. Mycol. J.* **2022**, *63*, 59–64. [CrossRef]
119. Seyedjavadi, S.S.; Bagheri, P.; Nasiri, M.J.; Razzaghi-Abyaneh, M.; Goudarzi, M. Fungal Infection in Co-Infected Patients with COVID-19: An Overview of Case Reports/Case Series and Systematic Review. *Front. Microbiol.* **2022**, *13*, 888452. [CrossRef]
120. Swain, S.; Ray, A.; Sarda, R.; Vyas, S.; Singh, G.; Jorwal, P.; Kodan, P.; Khanna, P.; Xess, I.; Sinha, S.; et al. COVID-19-Associated Subacute Invasive Pulmonary Aspergillosis. *Mycoses* **2022**, *65*, 57–64. [CrossRef] [PubMed]
121. Hoenigl, M.; Seidel, D.; Sprute, R.; Cunha, C.; Oliverio, M.; Goldman, G.H.; Ibrahim, A.S.; Carvalho, A. COVID-19-Associated Fungal Infections. *Nat. Microbiol.* **2022**, *7*, 1127–1140. [CrossRef]
122. Reizine, F.; Pinceaux, K.; Lederlin, M.; Autier, B.; Guegan, H.; Gacouin, A.; Luque-Paz, D.; Boglione-Kerrien, C.; Bacle, A.; Le Daré, B.; et al. Influenza- and COVID-19-Associated Pulmonary Aspergillosis: Are the Pictures Different? *J. Fungi* **2021**, *7*, 388. [CrossRef] [PubMed]
123. Verweij, P.E.; Rijnders, B.J.A.; Brüggemann, R.J.M.; Azoulay, E.; Bassetti, M.; Blot, S.; Calandra, T.; Clancy, C.J.; Cornely, O.A.; Chiller, T.; et al. Review of Influenza-Associated Pulmonary Aspergillosis in ICU Patients and Proposal for a Case Definition: An Expert Opinion. *Intensive Care Med.* **2020**, *46*, 1524–1535. [CrossRef]
124. Waldeck, F.; Boroli, F.; Suh, N.; Wendel Garcia, P.D.; Flury, D.; Notter, J.; Iten, A.; Kaiser, L.; Schrenzel, J.; Boggian, K.; et al. Influenza-Associated Aspergillosis in Critically-Ill Patients-a Retrospective Bicentric Cohort Study. *Eur. J. Clin. Microbiol. Infect. Dis.* **2020**, *39*, 1915–1923. [CrossRef]
125. Waldeck, F.; Boroli, F.; Zingg, S.; Walti, L.N.; Wendel-Garcia, P.D.; Conen, A.; Pagani, J.-L.; Boggian, K.; Schnorf, M.; Siegemund, M.; et al. Higher Risk for Influenza-Associated Pulmonary Aspergillosis (IAPA) in Asthmatic Patients: A Swiss Multicenter Cohort Study on IAPA in Critically Ill Influenza Patients. *Influenza Other Respir. Viruses* **2023**, *17*, e13059. [CrossRef] [PubMed]
126. Seldeslachts, L.; Staels, F.; Gkountzinopoulou, M.; Jacobs, C.; Tielemans, B.; Vanhoffelen, E.; Reséndiz-Sharpe, A.; De Herdt, L.; Haughton, J.; Prezzemolo, T.; et al. Damping Excessive Viral-Induced IFN-γ Rescues the Impaired Anti-*Aspergillus* Host Immune Response in Influenza-Associated Pulmonary Aspergillosis. *eBioMedicine* **2024**, *108*, 105347. [CrossRef]
127. Ahmad, A.; Singh, R.B.; Nickolich, K.L.; Pilewski, M.J.; Ngeow, C.; Frempong-Manso, K.; Robinson, K.M. Restoration of Type 17 Immune Signaling Is Not Sufficient for Protection during Influenza-Associated Pulmonary Aspergillosis. *Front. Immunol.* **2025**, *16*. [CrossRef]

128. Pantaleón García, J.; Wurster, S.; Albert, N.D.; Bharadwaj, U.; Bhoda, K.; Kulkarni, V.K.; Ntita, M.; Rodriguez Carstens, M.; Burch-Eapen, M.; Covarrubias López, D.; et al. Immunotherapy with Nebulized Pattern Recognition Receptor Agonists Restores Severe Immune Paralysis and Improves Outcomes in Mice with Influenza-Associated Pulmonary Aspergillosis. *mBio* **2025**, *16*, e04061-24. [CrossRef] [PubMed]
129. Vanderbeke, L.; Janssen, N.A.F.; Bergmans, D.C.J.J.; Bourgeois, M.; Buil, J.B.; Debaveye, Y.; Depuydt, P.; Feys, S.; Hermans, G.; Hoiting, O.; et al. Posaconazole for Prevention of Invasive Pulmonary Aspergillosis in Critically Ill Influenza Patients (POSA-FLU): A Randomised, Open-Label, Proof-of-Concept Trial. *Intensive Care Med.* **2021**, *47*, 674–686. [CrossRef]
130. Azoulay, É.; Afessa, B. Diagnostic Criteria for Invasive Pulmonary Aspergillosis in Critically Ill Patients. *Am. J. Respir. Crit. Care Med.* **2012**, *186*, 8–10. [CrossRef]
131. Ascioglu, S.; Rex, J.H.; de Pauw, B.; Bennett, J.E.; Bille, J.; Crokaert, F.; Denning, D.W.; Donnelly, J.P.; Edwards, J.E.; Erjavec, Z.; et al. Defining Opportunistic Invasive Fungal Infections in Immunocompromised Patients with Cancer and Hematopoietic Stem Cell Transplants: An International Consensus. *Clin. Infect. Dis.* **2002**, *34*, 7–14. [CrossRef]
132. De Pauw, B.; Walsh, T.J.; Donnelly, J.P.; Stevens, D.A.; Edwards, J.E.; Calandra, T.; Pappas, P.G.; Maertens, J.; Lortholary, O.; Kauffman, C.A.; et al. Revised Definitions of Invasive Fungal Disease from the European Organization for Research and Treatment of Cancer/Invasive Fungal Infections Cooperative Group and the National Institute of Allergy and Infectious Diseases Mycoses Study Group (EORTC/MSG) Consensus Group. *Clin. Infect. Dis.* **2008**, *46*, 1813–1821. [CrossRef] [PubMed]
133. Jenks, J.D.; Nam, H.H.; Hoenigl, M. Invasive Aspergillosis in Critically Ill Patients: Review of Definitions and Diagnostic Approaches. *Mycoses* **2021**, *64*, 1002–1014. [CrossRef] [PubMed]
134. Liu, R.-T.; Chen, Y.; Li, S.; Wan, X.-X.; Weng, L.; Peng, J.-M.; Du, B. A Comparison of Diagnostic Criteria for Invasive Pulmonary Aspergillosis in Critically Ill Patients. *Infect. Dis. Ther.* **2023**, *12*, 1641–1653. [CrossRef]
135. Lass-Flörl, C.; Resch, G.; Nachbaur, D.; Mayr, A.; Gastl, G.; Auberger, J.; Bialek, R.; Freund, M.C. The Value of Computed Tomography-Guided Percutaneous Lung Biopsy for Diagnosis of Invasive Fungal Infection in Immunocompromised Patients. *Clin. Infect. Dis.* **2007**, *45*, e101–e104. [CrossRef]
136. Lass-Flörl, C. How to Make a Fast Diagnosis in Invasive Aspergillosis. *Med. Mycol.* **2019**, *57*, S155–S160. [CrossRef]
137. Desoubeaux, G.; Bailly, É.; Chandenier, J. Diagnosis of Invasive Pulmonary Aspergillosis: Updates and Recommendations. *Med. Mal. Infect.* **2014**, *44*, 89–101. [CrossRef]
138. Garg, A.; Bhalla, A.S.; Naranje, P.; Vyas, S.; Garg, M. Decoding the Guidelines of Invasive Pulmonary Aspergillosis in Critical Care Setting: Imaging Perspective. *Indian J. Radiol. Imaging* **2023**, *33*, 382–391. [CrossRef]
139. Casutt, A.; Lamoth, F.; Lortholary, O.; Prior, J.O.; Tonglet, A.; Manuel, O.; Bergeron, A.; Beigelman-Aubry, C. Atypical Imaging Patterns during Lung Invasive Mould Diseases: Lessons for Clinicians. *Eur. Respir. Rev.* **2023**, *32*, 230086. [CrossRef] [PubMed]
140. Ghosh, I.; Raina, V.; Kumar, L.; Sharma, A.; Bakhshi, S.; Iqbal, S. Serum Galactomannan Assay for Diagnosis of Probable Invasive Aspergillosis in Acute Leukemia and Hematopoietic Stem Cell Transplantation. *Indian. J. Med. Paediatr. Oncol.* **2013**, *34*, 74–79. [CrossRef]
141. Shin, D.W.; Cho, J.; Choi, K.S.; Lee, J.; Choi, Y.; Choi, S.J.; Kim, S.-A.; Moon, S.M.; Kim, E.S.; Kim, H.B.; et al. False-Positive Results of Galactomannan Assays in Patients Administered Glucose-Containing Solutions. *Sci. Rep.* **2024**, *14*, 2552. [CrossRef]
142. Lamoth, F.; Akan, H.; Andes, D.; Cruciani, M.; Marchetti, O.; Ostrosky-Zeichner, L.; Racil, Z.; Clancy, C.J. Assessment of the Role of 1,3-β-d-Glucan Testing for the Diagnosis of Invasive Fungal Infections in Adults. *Clin. Infect. Dis.* **2021**, *72*, S102–S108. [CrossRef] [PubMed]
143. White, P.L.; Posso, R.B.; Barnes, R.A. Analytical and Clinical Evaluation of the PathoNostics AsperGenius Assay for Detection of Invasive Aspergillosis and Resistance to Azole Antifungal Drugs during Testing of Serum Samples. *J. Clin. Microbiol.* **2015**, *53*, 2115–2121. [CrossRef]
144. Gonçalves, S.M.; Lagrou, K.; Rodrigues, C.S.; Campos, C.F.; Bernal-Martínez, L.; Rodrigues, F.; Silvestre, R.; Alcazar-Fuoli, L.; Maertens, J.A.; Cunha, C.; et al. Evaluation of Bronchoalveolar Lavage Fluid Cytokines as Biomarkers for Invasive Pulmonary Aspergillosis in At-Risk Patients. *Front. Microbiol.* **2017**, *8*, 2362. [CrossRef] [PubMed]
145. Cadena, J.; Thompson, G.R.; Patterson, T.F. Aspergillosis: Epidemiology, Diagnosis, and Treatment. *Infect. Dis. Clin. N. Am.* **2021**, *35*, 415–434. [CrossRef]
146. Orasch, T.; Prattes, J.; Faserl, K.; Eigl, S.; Düttmann, W.; Lindner, H.; Haas, H.; Hoenigl, M. Bronchoalveolar Lavage Triacetylfusarinine C (TAFC) Determination for Diagnosis of Invasive Pulmonary Aspergillosis in Patients with Hematological Malignancies. *J. Infect.* **2017**, *75*, 370–373. [CrossRef]
147. Koo, S.; Thomas, H.R.; Daniels, S.D.; Lynch, R.C.; Fortier, S.M.; Shea, M.M.; Rearden, P.; Comolli, J.C.; Baden, L.R.; Marty, F.M. A Breath Fungal Secondary Metabolite Signature to Diagnose Invasive Aspergillosis. *Clin. Infect. Dis.* **2014**, *59*, 1733–1740. [CrossRef]

148. Machata, S.; Müller, M.M.; Lehmann, R.; Sieber, P.; Panagiotou, G.; Carvalho, A.; Cunha, C.; Lagrou, K.; Maertens, J.; Slevogt, H.; et al. Proteome Analysis of Bronchoalveolar Lavage Fluids Reveals Host and Fungal Proteins Highly Expressed during Invasive Pulmonary Aspergillosis in Mice and Humans. *Virulence* **2020**, *11*, 1337–1351. [CrossRef] [PubMed]
149. Lestrade, P.P.A.; Meis, J.F.; Melchers, W.J.G.; Verweij, P.E. Triazole Resistance in Aspergillus Fumigatus: Recent Insights and Challenges for Patient Management. *Clin. Microbiol. Infect.* **2019**, *25*, 799–806. [CrossRef] [PubMed]
150. Patterson, T.F.; Boucher, H.W.; Herbrecht, R.; Denning, D.W.; Lortholary, O.; Ribaud, P.; Rubin, R.H.; Wingard, J.R.; DePauw, B.; Schlamm, H.T.; et al. Strategy of Following Voriconazole versus Amphotericin B Therapy with Other Licensed Antifungal Therapy for Primary Treatment of Invasive Aspergillosis: Impact of Other Therapies on Outcome. *Clin. Infect. Dis.* **2005**, *41*, 1448–1452. [CrossRef] [PubMed]
151. Maertens, J.A.; Raad, I.I.; Marr, K.A.; Patterson, T.F.; Kontoyiannis, D.P.; Cornely, O.A.; Bow, E.J.; Rahav, G.; Neofytos, D.; Aoun, M.; et al. Isavuconazole versus Voriconazole for Primary Treatment of Invasive Mould Disease Caused by Aspergillus and Other Filamentous Fungi (SECURE): A Phase 3, Randomised-Controlled, Non-Inferiority Trial. *Lancet* **2016**, *387*, 760–769. [CrossRef]
152. Maertens, J.A.; Rahav, G.; Lee, D.-G.; Ponce-de-León, A.; Ramírez Sánchez, I.C.; Klimko, N.; Sonet, A.; Haider, S.; Diego Vélez, J.; Raad, I.; et al. Posaconazole versus Voriconazole for Primary Treatment of Invasive Aspergillosis: A Phase 3, Randomised, Controlled, Non-Inferiority Trial. *Lancet* **2021**, *397*, 499–509. [CrossRef]
153. Sanford Guide AMT: Aspergillosis, Invasive. Available online: https://web.sanfordguide.com/en/sanford-guide-online/disease-clinical-condition/aspergillosis-invasive (accessed on 17 July 2025).
154. Walsh, T.J.; Raad, I.; Patterson, T.F.; Chandrasekar, P.; Donowitz, G.R.; Graybill, R.; Greene, R.E.; Hachem, R.; Hadley, S.; Herbrecht, R.; et al. Treatment of Invasive Aspergillosis with Posaconazole in Patients Who Are Refractory to or Intolerant of Conventional Therapy: An Externally Controlled Trial. *Clin. Infect. Dis.* **2007**, *44*, 2–12. [CrossRef]
155. Viscoli, C.; Herbrecht, R.; Akan, H.; Baila, L.; Sonet, A.; Gallamini, A.; Giagounidis, A.; Marchetti, O.; Martino, R.; Meert, L.; et al. An EORTC Phase II Study of Caspofungin as First-Line Therapy of Invasive Aspergillosis in Haematological Patients. *J. Antimicrob. Chemother.* **2009**, *64*, 1274–1281. [CrossRef]
156. Kontoyiannis, D.P.; Ratanatharathorn, V.; Young, J.-A.; Raymond, J.; Laverdière, M.; Denning, D.W.; Patterson, T.F.; Facklam, D.; Kovanda, L.; Arnold, L.; et al. Micafungin Alone or in Combination with Other Systemic Antifungal Therapies in Hematopoietic Stem Cell Transplant Recipients with Invasive Aspergillosis. *Transpl. Infect. Dis.* **2009**, *11*, 89–93. [CrossRef]
157. Maschmeyer, G.; Haas, A.; Cornely, O.A. Invasive Aspergillosis: Epidemiology, Diagnosis and Management in Immunocompromised Patients. *Drugs* **2007**, *67*, 1567–1601. [CrossRef]
158. De Francesco, M.A. Drug-Resistant Aspergillus Spp.: A Literature Review of Its Resistance Mechanisms and Its Prevalence in Europe. *Pathogens* **2023**, *12*, 1305. [CrossRef]
159. Fakhim, H.; Badali, H.; Dannaoui, E.; Nasirian, M.; Jahangiri, F.; Raei, M.; Vaseghi, N.; Ahmadikia, K.; Vaezi, A. Trends in the Prevalence of Amphotericin B-Resistance (AmBR) among Clinical Isolates of *Aspergillus* Species. *J. Med. Mycol.* **2022**, *32*, 101310. [CrossRef]
160. Kaya, A.; Kaya, S.Y.; Balkan, İ.İ.; Alkan, S.; Kurt, A.F.; Elverdi, T.; Ürkmez, S.; Öngören, Ş.; Aygün, G. Amphotericin B Resistant *Aspergillus* spp.: Report of Two Cases. *Infect. Dis. Clin. Microbiol.* **2024**, *6*, 343–348. [CrossRef]
161. Chowdhary, A.; Sharma, C.; Meis, J.F. Azole-Resistant Aspergillosis: Epidemiology, Molecular Mechanisms, and Treatment. *J. Infect. Dis.* **2017**, *216*, S436–S444. [CrossRef] [PubMed]
162. Wiederhold, N.P. Epidemiology and Prevalence of Azole-Resistant Aspergillus Fumigatus: What Is Our Understanding of the Situation? *Curr. Fungal Infect. Rep.* **2023**, *17*, 177–187. [CrossRef]
163. Verweij, P.E.; Lucas, J.A.; Arendrup, M.C.; Bowyer, P.; Brinkmann, A.J.F.; Denning, D.W.; Dyer, P.S.; Fisher, M.C.; Geenen, P.L.; Gisi, U.; et al. The One Health Problem of Azole Resistance in *Aspergillus fumigatus*: Current Insights and Future Research Agenda. *Fungal Biol. Rev.* **2020**, *34*, 202–214. [CrossRef]
164. Dickwella Widanage, M.C.; Gautam, I.; Sarkar, D.; Mentink-Vigier, F.; Vermaas, J.V.; Ding, S.-Y.; Lipton, A.S.; Fontaine, T.; Latgé, J.-P.; Wang, P.; et al. Adaptative Survival of Aspergillus Fumigatus to Echinocandins Arises from Cell Wall Remodeling beyond B−1,3-Glucan Synthesis Inhibition. *Nat. Commun.* **2024**, *15*, 6382. [CrossRef]
165. CDC Antimicrobial-Resistant Aspergillus. Available online: https://www.cdc.gov/aspergillosis/php/guidance/index.html (accessed on 17 July 2025).
166. Maertens, J.A.; Thompson, G.R.; Spec, A.; Donovan, F.M.; Hammond, S.P.; Bruns, A.H.W.; Rahav, G.; Shoham, S.; Johnson, R.; Rijnders, B.; et al. Olorofim for the Treatment of Invasive Fungal Diseases in Patients with Few or No Therapeutic Options: A Single-Arm, Open-Label, Phase 2b Study. *Lancet Infect. Dis.* **2025**. [CrossRef]

Journal of Fungi

Article

Nine-Year Surveillance of *Candida parapsilosis* Candidemia in a Cardiothoracic ICU: Insights into Mortality and Resistance

Caio Trevelin Sambo [1], Bianca Leal de Almeida [2], Gabriel Fialkovitz [3], Tiago Alexandre Cocio [4], Afonso Rafael da Silva Junior [5], Lumena Pereira Machado Siqueira [5], Isabela Cristina Oliveira Silva [1], Flavia Rossi [5], Thaís Guimarães [6], Rinaldo Focaccia Siciliano [3], Evangelina da Motta Pacheco de Araújo [5], Gilda Maria Barbaro Del Negro [4], Gil Benard [4], Tania Mara Varejão Strabelli [3] and Marcello Mihailenko Chaves Magri [1,5,*]

1 Division of Infectious and Parasitic Diseases, Hospital das Clínicas, Faculdade de Medicina, Universidade de São Paulo, São Paulo 05403-010, Brazil; caio.sambo@hc.fm.usp.br (C.T.S.); isabelaoliveira87@gmail.com (I.C.O.S.)
2 Hospital Infection Control and Infectious Diseases Service, Instituto do Cancer do Estado de Sao Paulo, Faculdade de Medicina, Universidade de São Paulo, São Paulo 01246-000, Brazil; bianca.almeida@hc.fm.usp.br
3 Infection Control Department, Instituto do Coração, Hospital das Clínicas, Universidade de São Paulo, São Paulo 05403-000, Brazil; gabriel.fialkovitz@hc.fm.usp.br (G.F.); rinaldo.siciliano@hc.fm.usp.br (R.F.S.); tania.strabelli@incor.usp.br (T.M.V.S.)
4 Laboratory of Medical Mycology-LIM53, Instituto de Medicina Tropical, Divisão de Dermatologia, Hospital das Clínicas, Faculdade de Medicina, Universidade de São Paulo, São Paulo 05403-900, Brazil; alexcocio@gmail.com (T.A.C.); gildadelnegro@gmail.com (G.M.B.D.N.); bengil60@gmail.com (G.B.)
5 Pathology Department, Laboratório de Microbiologia da Divisão de Laboratório Central, Hospital das Clínicas da Faculdade de Medicina USP (FMUSP), São Paulo 05403-010, Brazil; afonso.rafael@hc.fm.usp.br (A.R.d.S.J.); lumena.siqueira@hc.fm.usp.br (L.P.M.S.); f.rossi@hc.fm.usp.br (F.R.); evangelina.araujo@hc.fm.usp.br (E.d.M.P.d.A.)
6 Infection Control Department, Hospital das Clínicas, University of São Paulo, São Paulo 05403-010, Brazil; tais.guimaraes@hc.fm.usp.br
* Correspondence: marcello.magri@hc.fm.usp.br; Tel.: +55-11-2661-6530

Abstract: *Candida parapsilosis* has emerged as a prominent cause of nosocomial candidemia, particularly among critically ill patients. The increasing prevalence of fluconazole-resistant *C. parapsilosis* (FR-Cp) poses major therapeutic challenges, especially in resource-limited settings. We conducted a retrospective study of 144 patients with *C. parapsilosis* candidemia admitted to two post-surgical ICUs at a Brazilian tertiary cardiothoracic hospital between 2016 and August 2024. Demographic, clinical, microbiological, and therapeutic data were analyzed. Predictors of 30-day mortality were identified through multivariate logistic regression. The incidence density of *C. parapsilosis* candidemia ranged from 2.93 to 8.31 per 1000 hospitalizations. Fluconazole resistance was identified in 81% of isolates. Overall 30-day mortality was 55%. Independent risk factors for mortality included cardiopathy (OR: 19.36, $p = 0.006$), higher SOFA scores (OR: 1.54, $p = 0.003$), parenteral nutrition (OR: 29.77, $p = 0.013$), and dialysis (OR: 6.59, $p = 0.043$), while longer treatment duration was protective (OR: 0.81, $p < 0.001$). Fluconazole resistance was not independently associated with increased mortality. In this cohort of critically ill patients, *C. parapsilosis* candidemia was associated with high mortality and a high prevalence of fluconazole resistance. Clinical outcomes were mainly driven by host-related and therapeutic factors rather than antifungal resistance alone.

Keywords: fluconazole-resistant *C. parapsilosis*; tertiary cardiothoracic hospital; candidemia

1. Introduction

Over the past decades hospital fungal infections have significantly increased, with *Candida* spp. accounting for approximately 80% of nosocomial fungal infections [1]. *C. albicans* has historically been the most prevalent species in nosocomial infections, with regional variations in prevalence rates [2,3]. However, non-*albicans Candida* species, particularly *C. parapsilosis*, have emerged as significant pathogens over recent decades, becoming a leading cause of candidemia worldwide [4]. In Brazil, studies at major hospitals, such as the Hospital das Clínicas of the University of São Paulo, have documented the evolving epidemiology of *C. parapsilosis*, highlighting its increasing incidence and its ability to form biofilms on medical devices, which is often linked to healthcare-associated transmission [5,6].

The emergence of fluconazole-resistant *C. parapsilosis* (FR-Cp) strains has become a major global concern. Although historically considered susceptible to azoles, increasing resistance has been reported in recent years in several regions, including Latin America, southern Europe, and Asia. High resistance rates have been described in single-center studies from Brazil (67.9%), South Africa (78%), Mexico (54%), Turkey (26.4%), Italy (33%), France (9.2%), and, more recently, Spain, where a sharp rise from 3.8% in 2019 to 29.1% in 2021 was documented. Similar trends have also been observed in other Mediterranean countries such as Greece. Resistance is mainly driven by *ERG11* gene substitutions, particularly Y132F, associated with clonal transmission and persistent hospital outbreaks [7–10]. A meta-analysis estimated a global fluconazole resistance prevalence of 15.2%, underscoring the emerging threat posed by FR-Cp [8].

Despite the growing recognition of *C. parapsilosis* as the second or third most prevalent cause of candidemia in Brazilian hospitals, data on its epidemiology, resistance patterns, and outcomes in critically ill cardiac patients remain limited. Specialized cardiologic ICUs face unique challenges, including high surgical complexity, frequent use of invasive devices, and prolonged hospital stays, which may contribute to increased fungal burden and antifungal resistance. This study retrospectively investigates candidemia caused by *C. parapsilosis* in two post-surgical ICUs at a major Brazilian tertiary cardiothoracic center.

2. Materials and Methods

This study aims to conduct a retrospective analysis of candidemia cases caused by *C. parapsilosis* in cardiothoracic patients in two post-surgical Intensive Care Units (ICU) at the Instituto do Coração do Hospital das Clínicas da Faculdade de Medicina da Universidade de São Paulo, a Brazilian tertiary care cardiothoracic hospital performing approximately 2000 cardiac surgeries annually. Ethical approval for the study was obtained from the local ethics committee (approval date: 3 July 2025; approval file: 7.686.849). The informed consent was waived because this study was retrospective, with the review of medical records.

Data collection was conducted through an active surveillance in a database maintained by the local hospital infection control service of the Heart Institute, which consolidates the microbiological results of infections recorded within the service. All patients with a positive blood culture for *C. parapsilosis* were included in the analysis. None of these patients were excluded on the final analysis. Candidemia was defined as the isolation of *C. parapsilosis* from blood culture, with only the first episode considered for analysis. A second episode was defined as a new positive blood culture occurring after a period of at least 30 days from the first episode [11]. Persistent candidemia was defined as positive blood cultures for *Candida* species lasting for $\geq$5 days [12]. Effective therapy was defined as the administration of a systemic antifungal agent with in vitro activity against *C. parapsilosis*, receiving at least 48 h of antifungal treatment. Optimal therapy was defined as the combination of an echinocandin and early catheter removal, the latter considered when the catheter was removed within ≤2 days after the candidemia diagnosis [13]. Prior antibiotic use was defined as the administration of broad-spectrum antibiotics, including carbapenems, ceftazidime-avibactam, polymyxins, aminoglycosides, glycopeptides, or linezolid. Prior antifungal use was defined as the administration of any azoles, echinocandins, or amphotericin. Both were considered in the last 30 days before candidemia.

This study comprehensively analyzed variables to characterize patients with candidemia caused by *C. parapsilosis* and to evaluate clinical outcomes in this population. Demographic data included patient age, reported as the mean ± standard deviation, and gender distribution, presented as the proportion of male patients. Clinical characteristics encompassed comorbidities such as cardiac disease, lung disease, diabetes mellitus (DM), chronic kidney disease, and immunosuppression, all documented if present at the time of hospital admission. Additionally, cases of concurrent COVID-19 infection during hospitalization were recorded. The clinical course was further characterized by complications associated with candidemia, including deep-seated infections, endophthalmitis, endocarditis, sternal osteomyelitis, and intra-abdominal infections. The study also evaluated the use of mechanical ventilation, vasoactive drug therapy, dialysis, and parenteral nutrition. Dialysis, total parenteral nutrition (TPN), and mechanical ventilation were considered if patients underwent these interventions up to the date of candidemia diagnosis. Sequential Organ Failure Assessment (SOFA) score and the use of vasoactive drugs were assessed on the date of candidemia diagnosis. Key laboratory findings included platelet count, creatinine levels, bilirubin levels, leukocyte count, neutrophil count, and C-reactive protein (CRP) levels within a 7-day window before or after the day of candidemia diagnosis. Echocardiography and fundoscopy were considered for analysis only if performed at least 72 h after the date of candidemia diagnosis. The study evaluated antifungal therapies prescribed, including fluconazole, echinocandins, and amphotericin B, as well as the adequacy of treatment.

The collected samples were sent to the Microbiology Laboratory of the Central Laboratory Division, Hospital das Clínicas, Faculdade de Medicina, University of São Paulo. Positive cultures were screened on chromogenic medium (ChromID *Candida*, bioMérieux, Marcy L'Étoile, France), and species identification was carried out by Matrix Assisted Laser Desorption Ionization Time-of-Flight mass spectrometry (MALDI-TOF MS) (Vitek MS, bioMérieux, Marcy-l'Étoile, France), and to FLC susceptibility testing according to Clinical and Laboratory Standards Institute guideline M44 [14]. Etest (bioMérieux, Marcy-l'Étoile, France) on RPMI 1640 agar was used to determine fluconazole MICs, and results were interpreted according to CLSI M27 (5th ed., 2022) and CLSI M60 (3rd ed., 2023) guidelines [15,16]. Microbiological data focused on the identification of *Candida* species in clinical cultures, with particular emphasis on *C. parapsilosis* and its resistance to fluconazole. The primary outcome of interest was the 30-day crude mortality rate, expressed as the percentage of deaths within 30 days of candidemia diagnosis.

A total of 17 *C. parapsilosis* isolates were selected for molecular analysis based on their availability and preservation in our laboratory strain collection, where representative isolates are routinely stored at −80 °C in glycerol stocks. Molecular analyses were performed at the Laboratory of Medical Mycology (LIM-53), Institute of Tropical Medicine/Division of Dermatology, Hospital das Clínicas, Faculdade de Medicina, University of São Paulo. *C. parapsilosis* sensu stricto (s.s.) isolates were genotyped by microsatellite analysis through amplification of seven different loci with primers previously described [17]. PCR products were separated on 3% agarose gel, stained with GelRed™ (Biotium, Fremont, CA, USA), and visualized with the UVITEC gel documentation system (Cleaver Scientific, Rugby, Warks, UK). For phylogenetic analysis, a .tif file generated by the photodocumentation system was used as input data for the PyEph 1.4 software (PyEph Project, 2025). Clustering was performed by the unweighted pair group method with arithmetic mean (UPGMA). The analysis also included the *C. parapsilosis* s.s. C935 belonging to the outbreak-related cluster described elsewhere [9]. In addition, the *C. parapsilosis* s.s. reference strain ATCC 22019 and two *C.parapsilosis* s.s. isolates (4E and 14D) unrelated to the outbreak cluster were also included to obtain the dendrogram.

The incidence density of candidemia (IDC) was calculated by dividing the number of candidemia cases recorded each year by the total number of hospitalizations in the same period, expressed as cases per 1000 hospitalizations. To test the association between categorical variables and hospital 30-day mortality, we used the $\chi 2$ test, Student's *t*-test, or Fisher's exact test when appropriate, describing results by frequency and percentage. Continuous variables are presented as mean and standard deviation and were compared using the Mann–Whitney U test. For multivariate analysis, we constructed a binary logistic regression model. Variables showing $p < 0.1$ in the univariate analysis were selected for inclusion; variables not meeting the prerequisites for logistic regression were excluded from the model. Values of $p < 0.05$ indicate statistical significance. Software used is IBM Corp. Released 2024. IBM SPSS Statistics for IOS, Version 30.0.0.0. Armonk, NY, USA: IBM Corp.

3. Results

This retrospective study included 144 cases of *C. parapsilosis* candidemia diagnosed in critically ill patients admitted to two cardiothoracic intensive care units of a Brazilian tertiary care hospital between 2016 and August 2024. The median age of the cohort was 51.1 years, with a predominance of male patients (54.1%). Fluconazole resistance was identified in 81% of *C. parapsilosis* isolates. The crude 30-day mortality rate was 55%.

The dendrogram analysis based on microsatellite profiles revealed four distinct clusters (Figure 1). One cluster consisting of all *C. parapsilosis* s.s. isolates evaluated in this study, indicating that they formed a single clonal lineage. The second cluster included the reference strain ATCC 22019, representing a distinct lineage with no epidemiological relationship to the studied isolates. In addition, a third and fourth clusters were formed by isolates 4E and 14D, characterizing lineages also unrelated to the studied isolates. These results demonstrate that the isolates were restricted to a single dominant *C. parapsilosis* s. s. lineage, genetically close to the C935 isolate, while the other clusters represented independent reference genotypes.

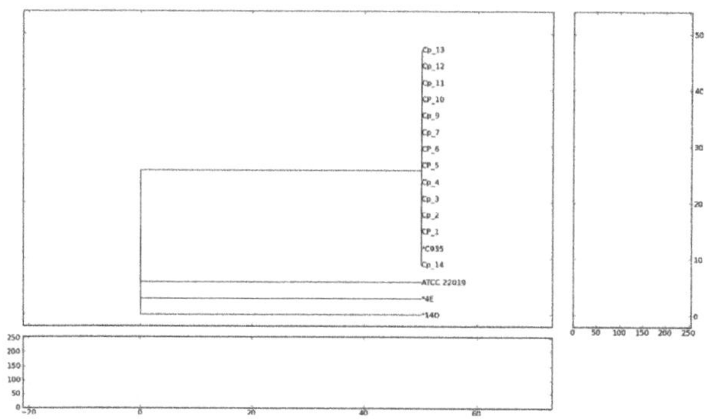

Figure 1. Dendrogram of *C. parapsilosis* s.s. isolates based on microsatellite profiles using the UPGMA method. Dendrogram showing the genetic relationship among the *C. parapsilosis* s.s. isolates, * 935C, 4E, 14D and the reference strain ATCC 22019, constructed using concatenated allele profiles from seven microsatellite loci and clustered by the UPGMA method.

The IDC increased from 2.93 cases per 1000 hospitalizations in 2016 to a peak of 8.31 in 2020, followed by a gradual decline to 6.03 in 2021, 5.17 in 2022, and 5.52 in 2023 (Figure 2A). As of August 2024, the IDC was 5.21 cases per 1000 hospitalizations. *C. parapsilosis* accounted for the majority of candidemia episodes, comprising 144 out of 237 cases (60.8%). Annual frequencies of *C. parapsilosis* ranged from 9 cases in 2016 to a peak of 22 cases in 2021, declining to 8 cases in 2024 (up to August). The distribution of other *Candida* spp. during the study period was as follows: *C. albicans* accounted for 34 cases (14.3%), *C. tropicalis* for 32 cases (13.5%), *C. glabrata* for 11 cases (4.6%), *C. dubliniensis* for 6 cases (2.5%), *C. guilliermondii* for 3 cases (1.3%), *C. krusei* for 4 cases (1.7%), *C. haemulonii* for 2 cases (0.8%), and *C. lusitaniae* for 1 case (0.5%). The temporal trends of all *Candida* spp. are summarized in Figure 2B.

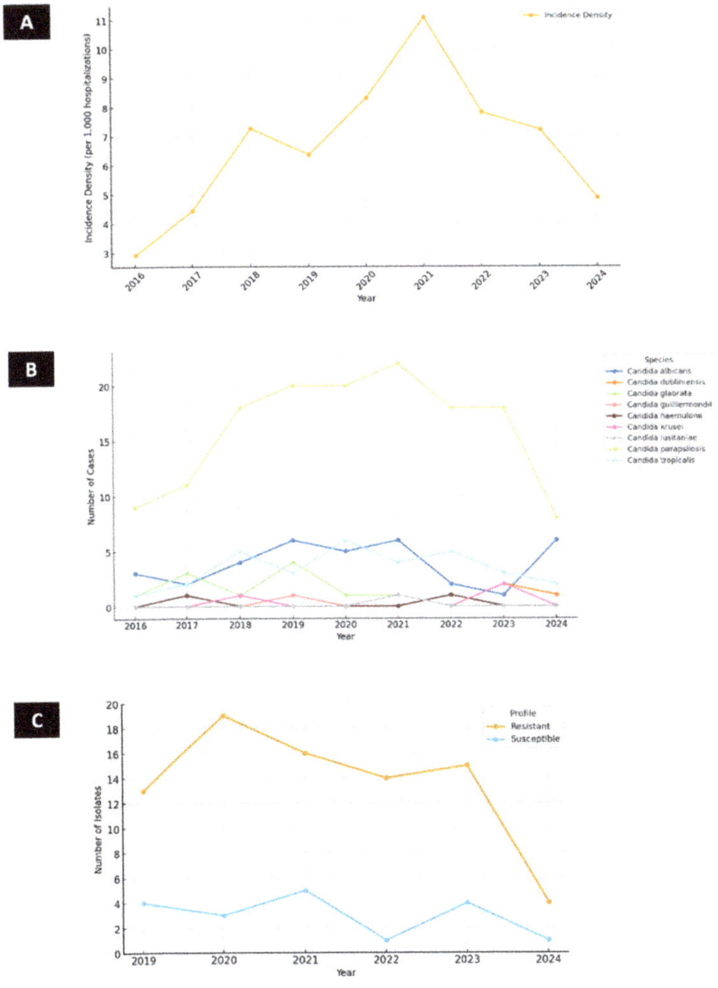

Figure 2. Temporal trends of candidemia episodes and fluconazole resistance in a cardiothoracic ICU (2016–2024). (**A**) Incidence density of candidemia episodes per 1000 hospitalizations over the study period. (**B**) Annual distribution of Candida species causing bloodstream infections. (**C**) Annual number of *C. parapsilosis* bloodstream isolates according to fluconazole susceptibility profile (resistant vs. susceptible).

The univariate analysis revealed that patients who died were older (63.1 ± 16.8 years) compared to survivors (58.1 ± 19.3 years), ($p = 0.016$, OR: 1.02 [99% CI: 1.00–1.03]). Gender did not show a significant association with mortality, as 70% of deceased patients were male, compared to 54.1% overall ($p = 0.575$, OR: 0.91 [99% CI: 0.68–1.23]). The length of stay in the ICU was notably shorter among patients who died (mean 52 days) compared to those who survived (mean 91 days), a highly significant finding ($p < 0.001$). Patient demographics, comorbidities, and clinical features by 30-day outcome are shown in Table 1.

Table 1. Univariate Analysis of Demographic, Comorbidity, and Clinical Variables Associated with 30-Day Mortality in ICU Patients with *Candida parapsilosis* Candidemia.

Variable	All Patients (144)	Alive (64)	Dead (80)	p	OR (IC 99%)
Age (mean, ST; years)	51.1 ± 22.6	58.1 ± 19.3	63.1 ± 16.8	0.360	1.02 (1.003–1.03)
Male	78 (54.1%)	33 (30%)	45 (70%)	0.575	0.91 (0.68–1.23)
Intensive Care Unit Admission	144 (100%)	64 (100%)	80 (100%)	-	-
Length of Stay	89 ± 62.8	91	52	<0.01	-
Comorbidities					
Cardiac disease	109 (75%)	43 (67%)	66 (82%)	0.003	2.30 (1.05–5.01)
Imunossupression	31 (21%)	14 (21%)	17 (21%)	0.928	0.96 (0.43–2.14)
Lung disease	17 (11%)	8 (12%)	9 (11%)	0.817	0.88 (0.32–2.44)
Diabetes mellitus	38 (26%)	15 (23%)	23 (28%)	0.472	1.31 (0.62–2.80)
Chronic Kidney disease	19 (13%)	6 (9%)	13 (16%)	0.226	1.87 (0.67–5.25)
COVID-19	13 (9%)	5 (7%)	8 (10%)	0.649	1.31 (0.40–4.22)
Surgery	132 (91%)	59 (92%)	73 (91%)	0.840	0.88 (0.26–2.92)
Complications					
Deep Site Infection (n=141)	31 (22%)	15 (23%)	16 (20%)	0.638	1.09 (0.74–1.59)
Endophthalmitis (n = 24)	1 (4%)	0	1 (4.3)	-	-
Endocarditis	8 (5%)	5 (7%)	3 (3%)	0.290	0.46 (0.10–2.00)
Sternal osteomyelitis	15 (10%)	6 (9%)	9 (11%)	0.714	1.22 (0.41–3.64)
Intra-abdominal infection	2 (1%)	0	2 (2.5%)	0.203	0.54 (0.47–0.63)
Surgical Site infection	18 (12%)	10 (15%)	8 (10%)	0.310	0.60 (0.22–1.62)
Persistent Candidemia (n = 80)	26 (32%)	14 (21%)	12 (15%)	0.536	0.84 (0.49–1.43)
Clinical characteristics					
Fever	29 (20%)	15 (23%)	14 (17.5%)	0.68	0.82 (0.28–2.41)
Source control (n = 36)	36 (25%)	18 (50%)	18 (50%)	0.14	2.12 (1.48–3.03)
Dialysis	83 (57%)	30 (46%)	53 (66%)	0.019	2.22 (1.13–4.37)
Vasoactive drug	106 (73%)	37 (57%)	69 (86%)	<0.01	4.57 (2.04–10.25)
SOFA (n = 110)	7.6 ± 4.2	4.83 ± 3.3	9.59 ± 3.72	<0.01	-
Mean arterial pression (n = 128)	70.15 ± 18.7	70.4 ± 20.7	69.9 ±17.2	0.54	-
Mechanical Ventilation	83 (57%)	28 (43%)	55 (68%)	0.003	2.82 (1.42–5.60)
Bacteremia	38 (26%)	18 (28%)	20 (25%)	0.179	1.07 (0.76–1.51)
Parenteral Nutrition	25 (17%)	5 (7%)	20 (25%)	0.007	3.93 (1.38–11.17)

SOFA = Sequential Organ Failure Assessment.

Among comorbidities, cardiac disease was significantly associated with 30-day mortality ($p = 0.003$, OR: 2.30 [99% CI: 1.05–5.01]). Other comorbidities, including immunosuppression ($p = 0.928$, OR: 1.01 [99% CI: 0.71–1.45]), DM ($p = 0.517$, OR: 0.88 [99% CI: 0.65–1.21]), chronic kidney disease ($p = 0.226$, OR: 1.87 [99% CI: 0.67–5.25]), lung disease ($p = 0.817$, OR: 0.88 [99% CI: 0.32–2.80]) and COVID-19 ($p = 0.207$, OR: 0.89 [99% CI: 0.56–1.41]), did not show significant associations with mortality.

Endocarditis had an OR of 0.46 [99% CI: 0.10–2.00], not statistically significant ($p = 0.290$). Endophthalmitis occurred in only one patient who died, limiting statistical analysis.

Patients requiring vasoactive drugs had significantly higher odds of death ($p < 0.001$, OR: 4.57 [99% CI: 2.04–10.25]), as did those on mechanical ventilation ($p = 0.03$, OR: 2.82 [99% CI: 1.42–5.60]). Dialysis was more common in deceased patients ($p = 0.019$, OR: 2.22 [99% CI: 1.13–4.37]), and parenteral nutrition showed a strong association with mortality ($p = 0.007$, OR: 3.93 [99% CI: 1.38–11.71]).

There was no statistically significant association between fluconazole resistance and 30-day mortality ($p = 0.113$, OR: 1.07 [99% CI: 0.71–1.62]). The proportion of resistant isolates was similar among survivors (54%) and deceased patients (57%), suggesting that fluconazole resistance alone may not be a primary determinant of mortality in this cohort. Figure 2C shows the annual distribution of fluconazole-resistant and susceptible *C. parapsilosis* isolates throughout the study period. Prior exposures, microbiological findings, treatment variables, and laboratory results by outcome are summarized in Table 2.

Table 2. Univariate Analysis of Treatment, Microbiological, and Laboratory Factors Associated with 30-Day Mortality in ICU Patients with *Candida parapsilosis* Candidemia.

Variable	All Patients (144)	Alive (64)	Dead (80)	p	OR (IC 99%)
Prior drug exposure					
Prior Antibiotics	140 (97%)	62 (96%)	78 (97.5%)	0.821	1.25 (0.17–9.18)
Prior antifungal	71 (49%)	27 (42%)	44 (55%)	0.126	1.67 (0.86–3.25)
Microbiology					
C.parapsilosis Fluconazole-Resistant (n = 99)	81 (81%)	35 (54%)	46 (57%)	0.113	1.07 (0.71–1.62)
>1 Candida on the same blood culture	3 (2%)	1 (1,5%)	2 (2.5%)	0.684	1.35 (0.31–5.90)
>1 Candida during CP candiemia	8 (5%)	3 (4%)	5 (6%)	0.684	0.88 (0.19–7.45)
C.albicans	5 (3%)	2 (3%)	3 (3%)	0.839	1.20 (0.44–1.91)
C.tropicalis	5 (3%)	1 (1.5%)	4 (5%)	0.263	3.31 (0.36–30.42)
C.glabrata	1 (0.6%)	0	1 (1.25%)	0.369	0.55 (0.47–0.64)
Prior *C.parapsilosis* colonization	42 (29%)	17 (26%)	25 (31%)	0.378	0.90 (0.66–1.23)
Treatment					
Antifungal prescribed	124 (86%)	55 (85%)	69 (86%)	1.000	1.03 (0.20 -3.58)
Effective therapy	109 (75%)	52 (94%)	57 (82%)	0.043	1.53 (1.12–2.08)
Fluconazole (n = 124)	28 (22%)	13 (23%)	15 (21%)	0.063	1.05 (0.71–1.54)
Echinocandin (n = 124)	91 (73%)	39 (71%)	52 (75%)	0.311	0.90 (0.61–1.31)
Amphotericin (n = 124)	8 (6%)	3 (5.5%)	5 (7.2%)	0.163	0.83 (0.50 –1.54)
Optimal Therapy	33 (22%)	22 (14%)	11 (18%)	0.003	1.53 (1.12–2.08)
Catheter removal (n = 140)	77 (55%)	48 (62%)	29 (37%)	<0.01	2.06 (1.50–2.83)
Early catheter removal (n = 81)	37 (45%)	22 (59%)	14 (40%)	0.700	0.89 (0.51–1.55)
Laboratory Findings					
Platelets ($\times 10^3/\text{mm}^3$)	159.770 ± 117.469	210.265 ± 119.541	119.375 ± 99.317	<0.01	-
Creatinine (mg/dL)	1.85 ± 1.24	1.68 ± 1.24	1.98 ± 1.24	0.038	-
Bilirrubins (n = 111)	2.54 ± 4.3	0.77 ± 0.54	3.75 ± 5.24	<0.01	-
Leucocytes	13.571 ± 8319	13.10 ± 7035	13.948 ± 9246	0.887	-
Neutrophils	10.668 ± 7095	10.474 ± 6475	10.823 ± 7593	0.963	-
PCR [(mg/L (n = 142)]	127 ± 75.8	103.6 ± 80.69	147.45 ±93.98	0.156	-
Time frames					
Days between candidemia and death	31.2 ± 32.1	56.7 ± 32.7	10.9 ± 8.14	<0.01	-
Days between hospitalization and candidemia	57.8 ± 49.3	55.08 ± 49.7	59.9 ± 49.2	0.604	-
Days between surgery and candidemia (n = 133)	28.7 ± 102.3	30.6 ± 49.2	27.2 ± 130.4	0.013	-
Days between candidemia and treatment (n = 86)	11.69 ± 78.56	2.32 ± 1.77	14.34 ± 88.97	0.371	-
Days between candidemia and first negative blood culture (n = 72)	18.7 ± 62	25.6 ± 74.4	4.17 ± 2.61	0.004	-
Days between candidemia and catheter removal (n = 79)	5.4 ± 34	1.5 ± 13.6	11.7 ± 52.9	0.406	-
Total treatment duration (n = 123)	22 ± 20.7	33 ± 25.4	13.3 ± 9.7	<0.01	-
30-Day crude mortality	80 (55%)	-	-	-	-

SOFA = Sequential Organ Failure Assessment. Colonization was defined as the presence of *Candida* species in cultures from non-sterile sites without signs or symptoms of active infection. PCR = Reactive C protein.

Effective antifungal therapy was a protective factor, with survivors receiving it in 94% of cases compared to 82% among deceased patients ($p = 0.043$, OR: 1.53 [99% CI: 1.12–2.08]). Timely catheter removal was also significantly protective ($p < 0.001$, OR: 2.06 [99% CI: 1.50–2.83]), although early removal showed no significant association ($p = 0.700$, OR: 0.89 [99%

CI: 0.51–1.55]). The choice of antifungal treatment (fluconazole, echinocandins, or amphotericin B) did not show significant associations with mortality ($p > 0.05$).

Platelet counts were significantly lower in patients who died (119,375 ± 99,317) compared to survivors (210,265 ± 119,541; $p < 0.001$). Bilirubin levels were markedly higher in deceased patients (3.75 ± 5.24 mg/dL) than in survivors (0.77 ± 0.54 mg/dL; $p < 0.001$).

The time between candidemia diagnosis and death was significantly shorter for deceased patients (10.9 ± 8.14 days) compared to survivors (56.7 ± 32.7 days; $p < 0.001$). Conversely, total treatment duration was significantly longer for survivors (33 ± 25.4 days) compared to deceased patients (13.3 ± 9.7 days; $p < 0.001$). The time between surgery and candidemia was shorter among deceased patients ($p = 0.013$), and the time to first negative blood culture was significantly shorter for deceased patients ($p = 0.004$).

Multivariate logistic regression identified five independent factors associated with 30-day mortality (Table 3). Pre-existing cardiopathy (OR = 19.356, 95% CI: 2.293–163.3, $p = 0.006$), higher SOFA scores (OR = 1.54, 95% CI: 1.158–2.049, $p = 0.003$), parenteral nutrition (OR = 29.767, 95% CI: 2.032–436.0, $p = 0.013$), and dialysis (OR = 6.59, 95% CI: 1.059–41.162, $p = 0.043$) were associated with increased mortality risk. Conversely, longer treatment duration was protective (OR = 0.807, 95% CI: 0.717–0.907, $p < 0.001$). These findings highlight the interplay of clinical factors influencing outcomes in ICU patients with candidemia.

Table 3. Multivariate analysis for mortality in 30 days.

Variable	p	OR	CI 95%
Cardiac disease	0.006	19.356	2.293–163.3
SOFA	0.003	1.54	1.158–2.049
Parenteral nutrition	0.013	29.767	2.032–436.0
Dialysis	0.043	6.59	1.059–41.162
Total treatment duration	<0.001	0.807	0.717–0.907

SOFA = Sequential Organ Failure Assessment.

4. Discussion

The results of this study provide valuable insights into the epidemiology and clinical outcomes of candidemia caused by *C. parapsilosis* in cardiologic critically ill patients, particularly in a resource-limited setting in Brazil. The study revealed a marked increase in incidence density during the COVID-19 pandemic, peaking in 2020, followed by a subsequent decline. Furthermore, the predominance of *C. parapsilosis* as the main pathogen and the alarming fluconazole resistance rate of 81% observed in this study highlight the unique challenges faced in managing candidemia in this region.

The incidence of candidemia varies significantly across different geographic regions, healthcare settings, and patient populations. In this study, the incidence density of candidemia caused by *C. parapsilosis* ranged from 2.93 cases per 1000 hospitalizations in 2016 to a peak of 8.31 cases per 1000 hospitalizations in 2020, with a subsequent decline in the following years. These findings contrast with a study from the same center [6], which reported incidence rates of up to 2.3 per 1000 hospital admissions and up to 1.3 per 1000 admissions in a tertiary care hospital in Rio de Janeiro [18]. Comparatively, data from developed countries show similar incidence rates when overall candidemia is compared to *C. parapsilosis* candidemia in this study. For instance, a multicenter study in Europe reported a cumulative incidence of 7.07 episodes per 1000 ICU admissions, with candidemia accounting for 5.52 episodes per 1000 admissions [19]. The predominance of *C. parapsilosis* in this study differs with trends observed in Brazil and other middle-income countries. While in this study *C. parapsilosis* accounted for 60% of candidemia cases in other studies *C. albicans* remains the main species accounting for 44%, 39% and 41.7% [6,20,21].

The temporal trends observed in this study, including the peak incidence during 2020, coincide with the impact of the COVID-19 pandemic. Increased ICU admissions, prolonged mechanical ventilation, and widespread use of immunosuppressive therapies likely contributed to the observed rise in cases [22,23]. This aligns with broader trends in Brazil, notably a 38% rise in candidemia incidence density reported by Jordana et al. [6] during 2020 compared to pre-pandemic years (2016–2019), reinforcing the pandemic's role in exacerbating fungal threats.

Regarding *C. parapsilosis* resistance, this study found 81% fluconazole resistant strains, a high burden even when compared to other reports in Brazil, where fluconazole resistance in *C. parapsilosis* has been documented to range from 6–11% and up to 26% in a study performed in Brazil's Midwest region particularly in tertiary care hospitals [24]. In 2022, Thomaz et al. described an interhospital transmission event between the Oncology Center of Hospital das Clínicas and the Cardiology Center of the present study during the COVID-19 pandemic in 2020 [23]. At that time, fluconazole-resistant strains accounted for 60% of isolates in the Cardiology Center, a profile that persisted throughout the years covered by this study, which reported an even higher resistance rate of 81%.

Despite the high prevalence of resistance, this study found no significant association between fluconazole resistance and 30-day mortality. In other study that evaluated FR-Cp during de COVID-19 pandemic showed 46.7% vs. 57.5% 30-day crude mortality between fluconazole resistant and susceptible groups with no statistical difference ($p = 0.312$) [22]. Also, in a French retrospective analysis of 78 invasive infections by *C. parapsilosis*, 30-day all-cause mortality between fluconazole susceptible and resistant strains were, respectively, 22.4% and 45.5% but also with no statistical significance ($p = 0.082$) [25]. An outbreak report of candidemia by fluconazole-resistant strains showed statistically significant difference regarding all-cause mortality (63.8% x 20%, $p = 0.008$) [26]. The lack of association in some studies may be attributed to the early use of alternative antifungals or to the predominance of fluconazole-resistant strains with lower virulence compared to other *Candida* species. Further studies are needed to evaluate the clinical and economic impact of FR-Cp candidemia.

In the present study, the overall 30-day mortality rate for candidemia was 55%, highlighting the significant burden of this infection in critically ill patients. This finding aligns with previous Brazilian studies, which report mortality rates up to 63.8% in tertiary centers [9]. A study in Turkey showed 50% mortality in candidemia by fluconazole-resistant strains [27]. A lower mortality rate from *C. parapsilosis* infection was noted in a prospective, population-based study among adults hospitalized in medical and surgical ICUs throughout Spain. The rate of death within 7 days for *C. parapsilosis* infection was 7%, versus 56% for *C. albicans*, which differs significantly from the present study [28].

Previous studies in Brazil that emphasize the role of organ dysfunction scores in predicting outcomes in overall candidemia. Araújo et al., 2024 [6] showed the following risk factors for 30-day mortality: advanced age, lung disease, heart disease, ICU stay, hemodialysis, use of corticosteroids, use of antibiotics, and mechanical ventilation. Colombo et al. 2012 showed association with higher 30-day mortality by univariate analysis: older age, period 1, higher APACHE II score, cancer, lung disease, renal failure, dialysis, mechanical ventilation, receipt of corticosteroids, no treatment for candidemia and treatment with deoxycholate amphotericin B. Infection due to *C. parapsilosis* though in that study was associated with lower mortality [21]. The only comorbidity associated with higher 30-day mortality was cardiac disease in both univariate and multivariate analysis. And this is also shown in other studies [6,21]. The multivariate analysis further refined the understanding of mortality predictors by identifying independent factors significantly associated with

30-day mortality. Among these, cardiac disease emerged as a key risk factor ($p = 0.006$, OR = 19.356), underscoring the vulnerability of patients with cardiovascular comorbidities. SOFA score remained a significant predictor in the multivariate model ($p = 0.003$, OR = 1.54), confirming the importance of systemic organ dysfunction as a determinant of outcomes. Other retrospective studies also highlighted how increased severity scores are associated with mortality, one which showed an increased risk when APACHE II score higher than 25 (OR 43.9) [28].

The need for parenteral nutrition and dialysis were also independently associated with mortality in the multivariate analysis, reinforcing their roles as indicators of critical illness. Parenteral nutrition has been identified as an independent risk factor for candidemia, including cases caused by *Candida parapsilosis*. The study by Poissy et al. demonstrated that total parenteral nutrition is an independent risk factor for candidemia in both ICU patients and those admitted to other hospital units [25]. Additionally, the study by Kutlu et al. also found a significant association between the use of parenteral nutrition and mortality in patients with candidemia [27].

The duration of antifungal treatment demonstrated a significant protective effect in our study ($p < 0.001$, OR = 0.807), highlighting the importance of prompt and adequate therapy. This finding aligns with previous research advocating for early and sustained antifungal treatment to improve outcomes in candidemia [21]. In our study, the failure to remove the catheter in the univariate analysis was associated with a higher risk of death ($p < 0.001$). Several studies have emphasized the critical role of early antifungal therapy and catheter removal in candidemia management. For instance, early appropriate antifungal treatment was shown to reduce mortality to 34.9%, and the combination of antifungal therapy with catheter removal within 48 h further decreased mortality to 18.9% (HR 0.34; 95% CI 0.16–0.70; $p = 0.03$) [28]. Similarly, a Spanish cohort study demonstrated that early antifungal therapy and catheter removal were independently associated with reduced early mortality (OR 0.27; 95% CI 0.08–0.91) [29]. These findings, consistent with our results, underscore the importance of timely interventions, particularly early antifungal therapy and catheter management, as essential components of candidemia treatment strategies to improve patient outcomes.

This study has several limitations that should be acknowledged. First, its retrospective nature may introduce biases related to data collection and the completeness of medical records. Missing data on certain variables, such as specific antifungal dosing, timing of catheter removal, and additional comorbidities, could have impacted the analysis of some risk factors and treatment outcomes. Second, the study was conducted in a single tertiary care hospital, which may limit the generalizability of the findings to other settings. Differences in healthcare infrastructure, patient demographics, and antifungal stewardship practices in other institutions and regions might result in variations in the epidemiology and outcomes of candidemia. Third, the high prevalence of fluconazole resistance observed in this study reflects the local epidemiological context and antifungal usage patterns, which might not be representative of other regions or countries. This limits the ability to extrapolate these findings to settings with lower rates of resistance or differing *Candida* species distributions. Another limitation is the relatively small sample size for certain subgroup analyses, such as the association between fluconazole resistance and mortality. Larger multicenter studies would provide a more robust assessment of the clinical impact of resistance and other factors. Another limitation of this study is that molecular analyses to detect mutations associated with azole resistance were not performed, as they are not part of our laboratory's routine workflow. Antifungal susceptibility testing was performed only for fluconazole, and other antifungal agents were not tested. Despite these limitations, the study provides valuable insights into the epidemiology, resistance patterns, and outcomes of candidemia in critically ill patients, particularly in the Brazilian healthcare context. These findings contribute to a better understanding of this challenging infection and highlight areas for improvement in clinical management and infection control.

5. Conclusions

This study highlights the significant burden of *C. parapsilosis* candidemia in critically ill patients, showing high rates of fluconazole resistance and mortality. While the high prevalence of fluconazole-resistant isolates is concerning, it was not independently associated with increased mortality, suggesting that early and appropriate antifungal treatment may mitigate its impact. The findings underscore the importance of continued antifungal stewardship, targeted infection control strategies, and early therapeutic interventions to improve outcomes in patients with candidemia. Furthermore, the study highlights the need for regional surveillance and multicenter collaborations to better understand the evolving epidemiology of *C. parapsilosis* candidemia, particularly in resource-limited settings like Brazil. Future prospective studies should aim to address the limitations identified, explore interventions to improve patient outcomes, and further investigate the clinical and microbiological factors contributing to mortality.

Author Contributions: C.T.S.: conceptualization, study design, data curation, formal analysis, writing—original draft, and writing—review and editing. B.L.d.A.: conceptualization, study design, data curation, formal analysis, writing—original draft, and writing—review and editing. G.F.: conceptualization, study design, and writing—original draft. T.A.C.: data curation and formal analysis. A.R.d.S.J.: data curation and formal analysis. L.P.M.S.: formal analysis. I.C.O.S.: visualization and data curation. F.R.: visualization and data curation. T.G.: writing—original draft and editing. R.F.S.: conceptualization and study design. E.d.M.P.d.A.: writing—original draft and editing. G.M.B.D.N.: conceptualization and formal analysis. G.B.: data curation and formal analysis. T.M.V.S.: conceptualization, writing—original draft and editing, and supervision. M.M.C.M.: visualization, study design, writing—original draft, writing—review and editing, and supervision. All authors have read and agreed to the published version of the manuscript.

Funding: This research received no external funding.

Institutional Review Board Statement: The study was conducted in accordance with the Declaration of Helsinki, and approved by the Ethics Committee of Faculdade de Medicina da Universidade de São Paulo, Brazil (CAAE 85382824.0.0000.0068, 3 July 2025).

Informed Consent Statement: Patient consent was waived due to the design of study (retrospective, based on analyses of clinical samples). All authors have read and agreed to the published version of the manuscript.

Data Availability Statement: Anonymized data may only be available upon well detailed and pertinent request due to privacy or ethical restrictions. Please contact the corresponding authors.

Conflicts of Interest: M.M.C.M. has received support for attending educational meetings from Knight and Mundipharma. The remaining authors declare no conflicts of interest.

References

1. da Silva Ruiz, L.; Montelli, A.C.; Sugizaki, M.F.; da Silva, E.G.; de Batista, G.C.; Moreira, D.; Paula, C.R. Outbreak of fungemia caused by *Candida parapsilosis* in a neonatal intensive care unit: Molecular investigation through microsatellite analysis. *Rev. Iberoam. Micol.* **2013**, *30*, 112–115. [CrossRef] [PubMed]
2. Pappas, P.G.; Lionakis, M.S.; Arendrup, M.C.; Ostrosky-Zeichner, L.; Kullberg, B.J. Invasive candidiasis. *Nat. Rev. Dis. Primers* **2018**, *4*, 18026. [CrossRef] [PubMed]

3. Hamburger, F.G.; Gales, A.C.; Colombo, A.L. Systematic review of candidemia in Brazil: Unlocking historical trends and challenges in conducting surveys in middle-income countries. *Mycopathologia* **2024**, *189*, 60. [CrossRef]
4. Delaloye, J.; Calandra, T. Invasive candidiasis as a cause of sepsis in the critically ill patient. *Virulence* **2014**, *5*, 161–169. [CrossRef] [PubMed]
5. Girão, E.; Levin, A.; Basso, M.; Gobara, S.; Gomes, L.; Medeiros, E.A.S.; Costa, S.F. Seven-year trend analysis of nosocomial candidemia and antifungal (fluconazole and caspofungin) use in Intensive Care Units at a Brazilian University Hospital. *Med. Mycol.* **2008**, *46*, 581–588. [CrossRef] [PubMed]
6. Araujo, J.M.; de Almeida Junior, J.N.; Magri, M.M.C.; Costa, S.F.; Guimarães, T. Epidemiological assessment and risk factors for mortality of bloodstream infections by *Candida* sp. and the impact of the COVID-19 pandemic era. *J. Fungi* **2024**, *10*, 268. [CrossRef]
7. Escribano, P.; Guinea, J. Fluconazole-resistant *Candida parapsilosis*: A new emerging threat in the fungi arena. *Front. Fungal Biol.* **2022**, *3*, 1010782. [CrossRef]
8. Yamin, D.; Akanmu, M.H.; Al Mutair, A.; Alhumaid, S.; Rabaan, A.A.; Hajissa, K. Global prevalence of antifungal-resistant *Candida parapsilosis*: A systematic review and meta-analysis. *Trop. Med. Infect. Dis.* **2022**, *7*, 188. [CrossRef]
9. Thomaz, D.Y.; de Almeida, J.N., Jr.; Sejas, O.N.E.; Del Negro, G.M.B.; Carvalho, G.O.M.H.; Gimenes, V.M.F.; de Souza, M.E.B.; Arastehfar, A.; Camargo, C.H.; Motta, A.L.; et al. Environmental clonal spread of azole-resistant *Candida parapsilosis* with Erg11-Y132F mutation causing a large candidemia outbreak in a Brazilian cancer referral center. *J. Fungi* **2021**, *7*, 259. [CrossRef]
10. Arastehfar, A.; Hilmioğlu-Polat, S.; Daneshnia, F.; Pan, W.; Hafez, A.; Fang, W.; Liao, W.; Şahbudak-Bal, Z.; Metin, D.Y.; De Almeida, J.N., Jr.; et al. Clonal candidemia outbreak by *Candida parapsilosis* carrying Y132F in Turkey: Evolution of a persisting challenge. *Front. Cell Infect. Microbiol.* **2021**, *11*, 676177.
11. Bassetti, M.; Giacobbe, D.R.; Agvald-Ohman, C.; Akova, M.; Alastruey-Izquierdo, A.; Arikan-Akdagli, S.; Azoulay, E.; Blot, S.; Cornely, O.A.; Cuenca-Estrella, M.; et al. Invasive Fungal Diseases in Adult Patients in Intensive Care Unit (FUNDICU): 2024 consensus definitions from ESGCIP, EFISG, ESICM, ECMM, MSGERC, ISAC, and ISHAM. *Intensive Care Med.* **2024**, *50*, 502–515. [CrossRef]
12. Pappas, P.G.; Kauffman, C.A.; Andes, D.R.; Clancy, C.J.; Marr, K.A.; Ostrosky-Zeichner, L.; Reboli, A.C.; Schuster, M.G.; Vazquez, J.A.; Walsh, T.J.; et al. Clinical Practice Guideline for the Management of Candidiasis: 2016 Update by the Infectious Diseases Society of America. *Clin. Infect. Dis.* **2016**, *62*, e1–e50. [CrossRef]
13. Agnelli, C.; Valerio, M.; Bouza, E.; Vena, A.; Guinea, J.; Martínez-Jiménez, M.d.C.; Marcos-Zambrano, L.J.; Escribano, P.; Muñoz, P. Persistent Candidemia in adults: Underlying causes and clinical significance in the antifungal stewardship era. *Eur. J. Clin. Microbiol. Infect. Dis.* **2019**, *38*, 607–614. [CrossRef]
14. CLSI. *Method for Antifungal Disk Diffusion Susceptibility Testing of Yeast*, 3rd ed.; CLSI Guideline M44; Clinical and Laboratory Standards Institute: Wayne, PA, USA, 2018.
15. CLSI. *Reference Method for Broth Dilution Antifungal Susceptibility Testing of Yeasts*, 4th ed.; CLSI Guidelines M27-Ed4; Clinical and Laboratory Standards Institute: Wayne, PA, USA, 2017.
16. CLSI. *Performance Standards for Antifungal Susceptibility Testing of Yeasts*, 2nd ed.; CLSI Guideline M60-Ed2; Clinical and Laboratory Standards Institute: Wayne, PA, USA, 2020.
17. Pulcrano, G.; Roscetto, E.; Iula, V.D.; Panellis, D.; Rossano, F.; Catania, M.R. MALDI-TOF mass spectrometry and microsatellite markers to evaluate *Candida parapsilosis* transmission in neonatal intensive care units. *Eur. J. Clin. Microbiol. Infect. Dis.* **2012**, *31*, 2919–2928. [CrossRef]
18. Braga, P.R.; Cruz, I.L.; Ortiz, I.; Barreiros, G.; Nouér, S.A.; Nucci, M. Secular trends of candidemia at a Brazilian tertiary care teaching hospital. *Braz. J. Infect. Dis.* **2018**, *22*, 273–277. [CrossRef]
19. Bassetti, M.; Giacobbe, D.R.; Vena, A.; Trucchi, C.; Ansaldi, F.; Antonelli, M.; Adamkova, V.; Alicino, C.; Almyroudi, M.-P.; Atchade, E.; et al. Incidence and outcome of invasive candidiasis in intensive care units (ICUs) in Europe: Results of the EUCANDICU project. *Crit. Care* **2019**, *23*, 219. [CrossRef]
20. Colombo, A.L.; Nucci, M.; Park, B.J.; Nouér, S.A.; Arthington-Skaggs, B.; da Matta, D.A.; Warnock, D.; Morgan, J. Brazilian Network Candidemia Study. Epidemiology of candidemia in Brazil: A nationwide sentinel surveillance of candidemia in eleven medical centers. *J. Clin. Microbiol.* **2006**, *44*, 2816–2823. [CrossRef] [PubMed]
21. Guimarães, T.; Nucci, M.; Mendonça, J.S.; Martinez, R.; Brito, L.R.; Silva, N.; Moretti, M.L.; Salomão, R.; Colombo, A.L. Epidemiology and predictors of a poor outcome in elderly patients with candidemia. *Int. J. Infect. Dis.* **2012**, *16*, e442–e447. [CrossRef]
22. Ramos-Martínez, A.; Pintos-Pascual, I.; Guinea, J.; Gutiérrez-Villanueva, A.; Gutiérrez-Abreu, E.; Díaz-García, J.; Asensio, Á.; Iranzo, R.; Sánchez-Romero, I.; Muñoz-Algarra, M.; et al. Impact of the COVID-19 pandemic on the clinical profile of candidemia and the incidence of fungemia due to fluconazole-resistant *Candida parapsilosis*. *J. Fungi* **2022**, *8*, 451. [CrossRef]

23. Thomaz, D.Y.; Del Negro, G.M.B.; Ribeiro, L.B.; da Silva, M.; Carvalho, G.O.M.H.; Camargo, C.H.; de Almeida, J.N.; Motta, A.L.; Siciliano, R.F.; Sejas, O.N.E.; et al. A Brazilian inter-hospital candidemia outbreak caused by fluconazole-resistant *Candida parapsilosis* in the COVID-19 era. *J. Fungi* **2022**, *8*, 100.
24. Mattos, K.; Rodrigues, L.C.; de Oliveira, K.M.P.; Diniz, P.F.; Marques, L.I.; Araujo, A.A.; Chang, M.R. Variability in the clinical distributions of *Candida* species and the emergence of azole-resistant non-*Candida albicans* species in public hospitals in the Midwest region of Brazil. *Rev. Soc. Bras. Med. Trop.* **2017**, *50*, 843–847. [CrossRef]
25. Poissy, J.; Damonti, L.; Bignon, A.; Khanna, N.; Von Kietzell, M.; Schrenzel, J.; Funginos, T.; Boggian, K.; Neofytos, D.; Vuotto, F.; et al. Risk factors for candidemia: A prospective matched case-control study. *Crit. Care* **2020**, *24*, 109. [CrossRef]
26. Hirai, Y.; Asahata, S.; Ainoda, Y.; Goto, A.; Fujita, T.; Totsuka, K. Nosocomial *Candida parapsilosis* candidemia: Risk factors, antifungal susceptibility and outcome. *J. Hosp. Infect.* **2014**, *87*, 54–58. [CrossRef]
27. Kutlu, M.; Sayın-Kutlu, S.; Alp-Çavuş, S.; Şirin, M.C.; Tülek, N.; Alkan, E.B.; Kaya, O.; Kutsoylu, O.E.; Şenol-Akar, Ş.; Turhan, Ö.; et al. Mortality-associated factors of candidemia: A multi-center prospective cohort in Turkey. *Eur. J. Clin. Microbiol. Infect. Dis.* **2022**, *41*, 597–607. [CrossRef]
28. Garnacho-Montero, J.; Díaz-Martín, A.; García-Cabrera, E.; Ruiz Pérez de Pipaón, M.; Hernández-Caballero, C.; Lepe-Jiménez, J.A. Impact on hospital mortality of catheter removal and adequate antifungal therapy in *Candida* spp. bloodstream infections. *J. Antimicrob. Chemother.* **2013**, *68*, 206–213. [CrossRef]
29. Puig-Asensio, M.; Pemán, J.; Zaragoza, R.; Garnacho-Montero, J.; Martín-Mazuelos, E.; Cuenca-Estrella, M.; Almirante, B. Impact of therapeutic strategies on the prognosis of candidemia in the ICU. *Crit. Care Med.* **2014**, *42*, 1423–1432. [CrossRef]

MDPI AG
Grosspeteranlage 5
4052 Basel
Switzerland
Tel.: +41 61 683 77 34

Journal of Fungi Editorial Office
E-mail: jof@mdpi.com
www.mdpi.com/journal/jof

www.ingramcontent.com/pod-product-compliance
Lightning Source LLC
Chambersburg PA
CBHW062037220926
51850CB00010B/2580
* 9 7 8 3 7 2 5 8 7 3 4 8 7 *